Synthesis Lectures on Information Concepts, Retrieval, and Services

Series Editor

Gary Marchionini, School of Information and Library Science,
The University of North Carolina at Chapel Hill, Chapel Hill, NC, USA

This series publishes short books on topics pertaining to information science and applications of technology to information discovery, production, distribution, and management. Potential topics include: data models, indexing theory and algorithms, classification, information architecture, information economics, privacy and identity, scholarly communication, bibliometrics and webometrics, personal information management, human information behavior, digital libraries, archives and preservation, cultural informatics, information retrieval evaluation, data fusion, relevance feedback, recommendation systems, question answering, natural language processing for retrieval, text summarization, multimedia retrieval, multilingual retrieval, and exploratory search.

Xiaojun Yuan · Dan Wu ·
DeeDee Bennett Gayle

Editors

Social Vulnerability to COVID-19

Impacts of Technology Adoption and Information Behavior

Editors
Xiaojun Yuan
College of Emergency Preparedness
Homeland Security, and Cybersecurity
University at Albany
State University of New York
Albany, NY, USA

Dan Wu
School of Information Management
Wuhan University
Wuhan, China

DeeDee Bennett Gayle
College of Emergency Preparedness
Homeland Security, and Cybersecurity
University at Albany
State University of New York
Albany, NY, USA

ISSN 1947-945X   ISSN 1947-9468   (electronic)
Synthesis Lectures on Information Concepts, Retrieval, and Services
ISBN 978-3-031-06899-7   ISBN 978-3-031-06897-3   (eBook)
https://doi.org/10.1007/978-3-031-06897-3

This Springer imprint is published by the registered company Springer Nature Switzerland AG
The registered company address is: Gewerbestrasse 11, 6330 Cham, Switzerland

Contents

Editors and Contributors

About the Editors

Xiaojun Yuan Ph.D., is an Associate Professor in the College of Emergency Preparedness, Homeland Security, and Cybersecurity at the University at Albany, State University of New York. Her research interests include both human–computer interaction and information retrieval, with a focus on user interface design and evaluation and human information behavior.

She has received various grants and contracts, including from the Institute of Museum and Library Services, SUNY seed grant, Initiatives for Women Program at the University at Albany, and New York State Education Department.

She published extensively in journals in information retrieval and human–computer interaction (JASIS&T, IP&M, Journal of Documentation, etc.), and conferences in computer science and information science (ACM SIGIR, ACM SIGCHI, ACM CHIIR, ASIS&T, etc.).

She received her Ph.D. from Rutgers University at the School of Communication and Information, and Ph.D. from Chinese Academy of Sciences in the Institute of Computing Technology. She received her M.S. in Statistics from Rutgers University and M.E. and B.E. in Computer Application from Xi'an University of Science & Technology in China. She serves as an Editorial Board Member of Aslib Journal of Information Management (AJIM) and a Board Member of the International Chinese Association of Human-Computer Interaction. She is a member of the Association for Information Science and Technology (ASIS&T), the Association for Computing Machinery (ACM), and the Institute of Electrical and Electronics Engineers (IEEE).

Dan Wu Ph.D., is a Professor in the School of Information Management at Wuhan University, a member of the Academic Committee of Wuhan University, and the Director of the Human-Computer Interaction and User Behavior Research Center. Her research areas include information organization and retrieval, user information behavior, human–computer interaction, and digital libraries.

She has secured research grants and contracts from the National Social Science Foundation of China (Major Program), the National Natural Science Foundation of China (NSFC), and the Humanities and Social Sciences Foundation of the Ministry of Education.

Her work is published in various journals (including the IP&M, JASIST, etc.). She has presented at several conferences related to information behavior, information retrieval, and digital library (SIGIR, CIKM, CSCW, CHIIR, etc.).

She received her Ph.D. from Peking University in Management. She serves as the Editor-in-Chief of Aslib Journal of Information Management and the Executive Editor of Data and Information Management. She also serves as a director at large of the ASIS&T. She is a member of the ACM Digital Library Committee, a member of the iSchool Data Science Curriculum Committee, and a member of the JCDL Steering Committee.

DeeDee Bennett Gayle Ph.D. is an Associate Professor in the College of Emergency Preparedness, Homeland Security, and Cybersecurity at the University at Albany, State University of New York. She broadly examines the influence and integration of advanced technologies on the practice of emergency management, and for use by vulnerable populations.

She has secured research grants and contracts, including from the National Science Foundation, the Federal Emergency Management Agency, and the Department of Homeland Security. Her work is published in various journals, and she has presented at several conferences related to emergency management, disability, wireless technology, and future studies.

She received her Ph.D. from Oklahoma State University in Fire and Emergency Management. She has a unique academic background having received both her M.S. in Public Policy and B.S. in Electrical Engineering from the Georgia Institute of Technology. She is an Advisory Board Member for the Institute for Diversity and Inclusion in Emergency Management (I-DIEM), a member of the Social Science Extreme Events Reconnaissance (SSEER) and Interdisciplinary Science Extreme Events Reconnaissance (ISEER), within the NSF-FUNDED CONVERGE initiative.

Contributors

Yuan Xiaojun University at Albany, State University of New York, Albany, NY, USA

Bennett Gayle DeeDee University at Albany, State University of New York, Albany, NY, USA

Dubois Elisabeth University at Albany, State University of New York, Albany, NY, USA

Huang Kun School of Government, Beijing Normal University, Beijing, China

Knight Thora University at Albany, State University of New York, Albany, NY, USA

LaForce Salimah Georgia Institute of Technology, Atlanta, GA, USA

Li Lei School of Government, Beijing Normal University, Beijing, China

Liang Shaobo School of Information Management, Wuhan University, Wuhan, China; Center for Studies of Human-Computer Interaction and User Behavior, Wuhan University, Wuhan, China

Luo Shi Chao School of Government, Beijing Normal University, Beijing, China

Ma Le School of Information Management, Wuhan University, Wuhan, China

Wang Xiao Yu School of Government, Beijing Normal University, Beijing, China

Wu Dan School of Information Management, Wuhan University, Wuhan, China

Introduction

DeeDee Bennett Gayle and Xiaojun Yuan

Abstract

This edited volume provides a unique account of the impact COVID-19 had on the use and adoption of technology among the most marginalized populations worldwide. Older adults were disproportionately at risk during the pandemic because of their infection and mortality rate, and at risk to social isolation due to lower levels of digital literacy, use, and adoption. However, other populations had high levels of social disruption during the pandemic, namely children (and families with children), who had to adapt to rapidly changing new school policies requiring them to use online technologies for education. The changes in course instruction put a strain on the children who may have difficulties learning virtually and the families who assisted the students with technology issues, provided broadband wireless access, and enables a quiet learning environment. This chapter provides a brief overview of the global impact of the pandemic, as well as the response measures which led to a dramatic increase in the worldwide reliance of communications technology. Additionally, this chapter includes a summary of the types of marginalized populations who may have had significant difficulty in use or adoption of technologies for employment, education, or other activities in daily life (such as doctor's visits and government services). Finally, a synopsis of the research studies included in this book are presented.

Keywords

COVID-19 • Technology • Marginalized populations • Book overview

D. Bennett Gayle (✉) · X. Yuan
University at Albany, State University of New York, Albany, NY, USA
e-mail: dmbennett@albany.edu

1.1   Global Impact of the Covid-19 Pandemic

In January of 2020, the world health organization published a comprehensive package providing guidance related to the outbreak of a new disease [34]. Authorities in China determined that they identified an outbreak caused by a novel coronavirus. That novel virus was named SARS-CoV2 and is frequently referred to as COVID-2019 or just coronavirus. By the 21st of January the United States (US) had its' first confirmed case of the virus [13]. On the 24th of January, France confirmed three cases of the novel virus [38]. And by January 29th the United Arab Emirates confirmed its first case [37]. The outbreak quickly turned into a pandemic. On January 30th the World Health Organization declared a Public Health Emergency of International Concern and on March 11th made the assessment that COVID-19 could be considered a pandemic [1, 5]. Initial advice from the WHO in February suggested that there should be a quarantine of individuals along with guidance on the conditions of the containment. And in March they offered interim guidance for preparedness, readiness, and response, among the public health measures was the suggestion to practice social distancing [36]. Maragakis [17]describes social distancing as a non-pharmaceutical measure in response to an infectious disease and means to stay home and away from others, the practice encourages the use of online and web-based technology for communication instead of in-person contact. Many nations and government bodies embraced this public health emergency response measure to quickly manage the pandemic.

However, by April 4th of 2020 WHO reported over 1 million cases worldwide [35]. In December of 2020 two new variants of the virus were detected in United Kingdom and South Africa and there were over 800, 000 daily cases worldwide. The first emergency use vaccination was not issued until December 31st, 2020, by the WHO [33]. Though several vaccines have been developed since, vaccinations have been slow and sporadic in many countries. For example, in the United States, early racial disparities were found with vaccinated individuals due to access, technology, and mistrust [25]. As of August 11th, 2021, there were nearly 204 million confirmed cases including over 4 million deaths due to COVID-19 reported to WHO, while nearly 4.4 million vaccine doses have been administered over the same time frame [39]. Therefore, many policies around social distancing remain. Figure 1.1 shows a timeline of the COVID-19 progression, several more detailed timelines may be found through the WHO, CDC, or AJMC.

The end of the pandemic is not in sight yet as a new Delta Variant of COVID-19 is more infectious leading to increased transmissions even among vaccinated individuals. New guidance on the Delta Variant began to emerge as early as July of 2021 even as the push to vaccinate individuals continued [3]. In the United States, which has highest total number of cases worldwide, many states are encouraging new or continues social

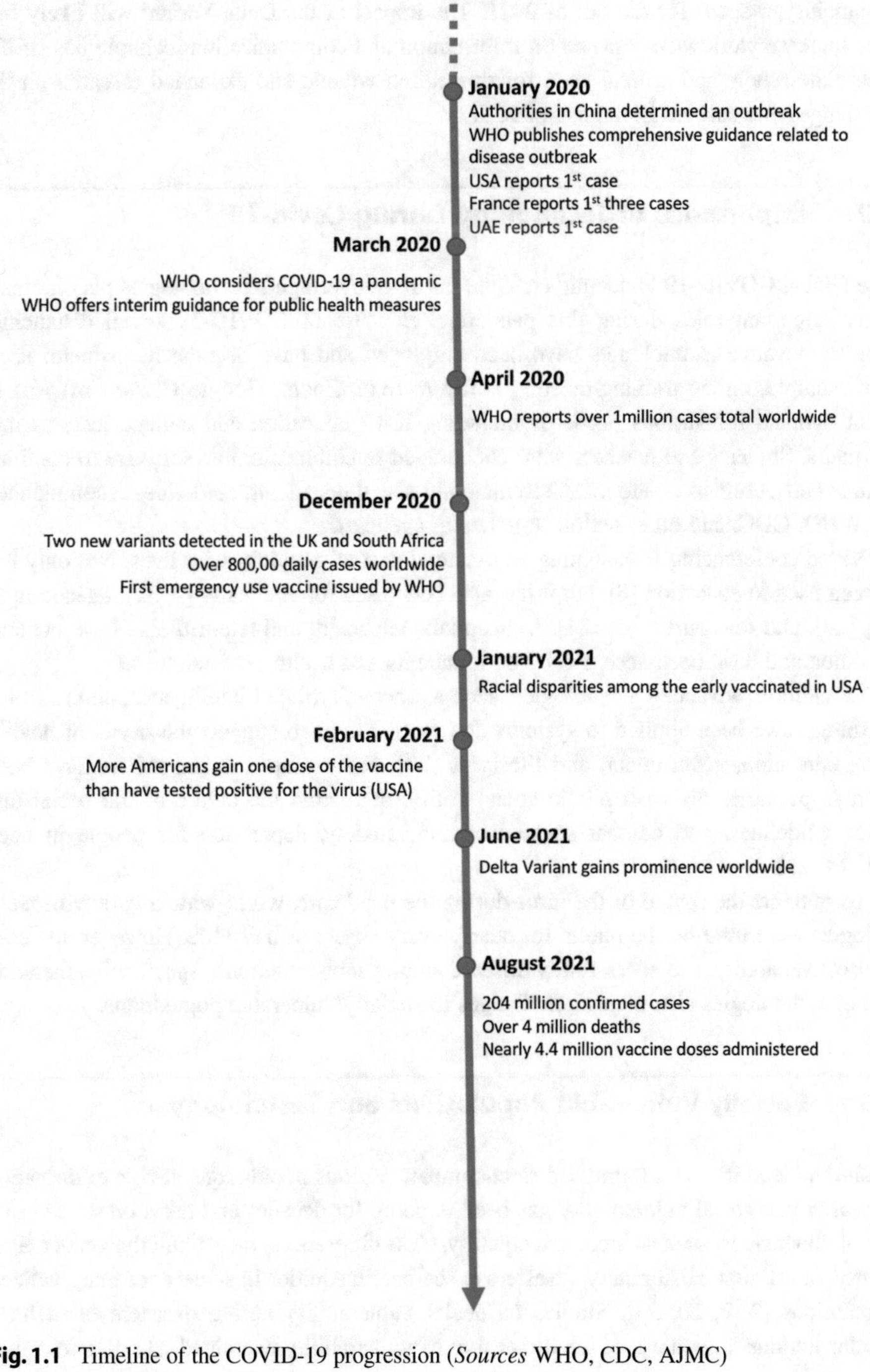

January 2020
Authorities in China determined an outbreak
WHO publishes comprehensive guidance related to
disease outbreak
USA reports 1st case
France reports 1st three cases
UAE reports 1st case

March 2020
WHO considers COVID-19 a pandemic
WHO offers interim guidance for public health measures

April 2020
WHO reports over 1million cases total worldwide

December 2020
Two new variants detected in the UK and South Africa
Over 800,00 daily cases worldwide
First emergency use vaccine issued by WHO

January 2021
Racial disparities among the early vaccinated in USA

February 2021
More Americans gain one dose of the vaccine
than have tested positive for the virus (USA)

June 2021
Delta Variant gains prominence worldwide

August 2021
204 million confirmed cases
Over 4 million deaths
Nearly 4.4 million vaccine doses administered

Fig. 1.1 Timeline of the COVID-19 progression (*Sources* WHO, CDC, AJMC)

distancing practices for the fall of 2021. The impact of the Delta Variant will likely further increase worldwide reliance on information and communication technologies (ICT). Thus, the timely and critical need for this edited volume and expanded research on the challenges and success in the use of ICT.

1.2   Importance of Technology During Covid-19

The Global COVID-19 Pandemic changed our lives overnight. Technologies play increasingly important roles during this pandemic. Because of COVID-19 social distancing, many innovative technologies have been employed and have become mainstream uses. For example, online training/meeting software (e.g., Zoom, Google Classroom) was in great demand for various purposes including K-12 education and college lectures and seminars. Students and teachers were encouraged to embrace online software to continue course curriculum as a safe means to maintain social distancing guidelines recommended by WHO, CDC, and other nation state health agencies.

Video conferencing is becoming an essential part of people's daily lives. Not only has it been used in education [8], but it has also been used for telehealth [4, 32], telecommuting [22], and for court trials [24]. In hospitals, telehealth and telemedicine have become common and have been accepted by both patients and health professionals.

In addition, advanced technologies, such as robots, artificial intelligence, and machine learning, have been applied to systems and mobile apps to support the needs of healthcare, education, government, and lifestyles [29]. For example, social robots have been used in public to disinfect public areas, notify, or remind the public of the social distance guidelines, and combat social isolation, anxiety, depression for people in need [11, 14, 21].

To mitigate the spread of the virus during the pandemic, world-wide a variety of technologies were used by the public for nearly every aspect of daily life. However, the ease of use, availability, and accessibility differed among subpopulations. Specifically, the wide use of technologies also present challenges to socially vulnerable populations.

1.3   Socially Vulnerable Populations and Technology

Social vulnerability is a term used to encompass various populations-at-risk to disasters. Research into social vulnerability has been ongoing for decades and takes on several different dimensions, such as income inequality, class differences, race/ethnicity, gender, age, ability, or religion. Historically, disasters exacerbate inequities in society for marginalized populations [2, 7, 20, 23]. Studies on social vulnerability during disasters often have similar findings regardless of the dimension of vulnerability examined, (1) disaster relief organizations should address the unique needs of (and underlying causes for) the vulnerable population [10, 28, 30], (2) vulnerable populations should be empowered during (and actively involved in) planning efforts and decision-making for their communities [2, 12, 18].

The social vulnerability perspective posits that marginalized populations may lack access to vital economic and social resources, possessing limited autonomy and power, and having low levels of social capital necessary for disaster response and recovery [20]. This perspective considers how social vulnerability is incorporated into disaster preparedness, response, recovery, and mitigation, beyond exposure [12]. For example, individuals experiencing homelessness, who tend to have the highest exposure during disasters, also have conflating factors that lead to their vulnerability such as social stigma, special medical needs, and chronic unemployment [27]. Additionally, access to vital resources prior to, during, and after disasters may increase the vulnerability of some marginalized populations. Resources are a broad term that may include a variety of needs, such as water [31], social support [15], finances [26], or information [19]. Often connected to the lack of access to information, is the connection to technology. For example, the increased reliance on social media for information during disasters has exposed a lack of access to broadband wireless or the appropriate technology due to power outages or system barriers such as the digital divide [9]. The digital divide is "the gap that exists between individuals who have access to modern information and communication technology and those who lack access [6]." This gap is often between the least and most marginalized in our societies. Though not a static or permanent position, digital divide has been exacerbated among socially vulnerable populations, and the situation will likely continue post pandemic. Lai and Widmar [16] revisited the digital divide situation in COVID-19 and reported that there existed a negative relation between rurality and internet speed, indicating the challenges presented in rural areas. Schools have tried to provide equitable educational access to all the students, but households who can barely afford the internet service suffer from maintaining sufficient Internet speeds. They summarized that "essential activities moved online, yet sufficient Internet is an essential public service that remains unattainable for many US households." It is worth mentioning that socially vulnerable populations not only suffer from digital divide, but also other aspects of their daily lives. Because of their disadvantageous status, many resources, technologies, or technological support were not available to them.

1.4 Organization of the Book

This book presents research studies relating to the access to technology during COVID by some socially vulnerable populations, including children, older adults, COVID-19 patients, and general marginalized populations. The book covers several technologies, as well, from artificial intelligence to telehealth and telecommunications.

In the paragraphs below, a brief synopsis of each chapter is provided to assist the reader. The authors of each chapter introduce the dimension of vulnerability explored and the impacts faced by a specific technology (or technologies). Chapters 2 and 3 provide unique literature reviews to set the stage for the overall impact beyond what is presented in the following empirical studies highlighted in Chaps. 4–6. Chapter 7 provides considerations which may impact privacy and security. The concluding chapter provides an overview of lessons learned and the path forward to reduce the digital divide, thereby reducing the vulnerability to the COVID-19 virus.

In Chap. 2, authors Bennett Gayle, Yuan, Dubois, and Knight present a research agenda for use, access, and adoption of information and communication technologies (ICT) during the pandemic among children, older adults, people with disabilities and other marginalized groups. The research agenda is based on a literature review, with articles investigated using the ecological systems framework.

In Chap. 3, authors Yuan, Bennett Gayle, Knight, and Dubois present a systematic literature review related to the use of artificial intelligence (AI) for socially vulnerable populations during the pandemic. Using three databases to review empirical studies, the authors find a significant gap in the literature regarding the impacts of AI on various subpopulations.

In Chap. 4, authors Wu and Ma examine the question-and-answer data (Q&A data) on the health information needs of COVID-19 patients under the "Baidu Zhidao" community, firstly clarifies the topic of health information needs of COVID-19 patients, and then further explores the law of changes in the topic of health information needs of COVID-19 patients over time, and finally analyzes users' social attributes and theme COVID-19 patients health information demand. Based on clarifying the content of the health information needs of COVID-19 patients, the above analysis can further provide a positive reference for the socialized Q&A community to improve the service level.

In Chap. 5, author Liang discusses the digital divide faced by the elderly. During the pandemic, older adults' lack of understanding of the epidemic information has resulted in poor self-protection, limited travel, and affected the purchase of life and medical items. In the chapter, the authors used a semi-structured interview of 25 older adults to analyze the difficulties of information acquisition and utilization of the elderly. The findings of that study can help solve the digital divide faced by the elderly in the epidemic, help older adults better obtain accurate information, and help them better protect their health.

In Chap. 6, authors Huang, Li, Luo, and Wang discuss disparities in accessing the public information during the pandemic by the most vulnerable. Public information service (PIS) is the premise and basis for people to protect their own health and safety and make scientific and reasonable decision-making. Thus, insight and understanding of difficulties of vulnerable groups in accessing PIS during the emergencies is important. Since the pandemic measures and recovery are different by country, the policies and measures adopted by the governments are considered in this study. Using China as an example, the authors focus on the following issues: (1) What difficulties do the vulnerable groups in

China encounter when they try to access PIS? (2) How to improve the PIS to better help vulnerable groups survive in the public health emergencies?

In Chap. 7, authors Knight, Yuan, Bennett Gayle and LaForce discuss the implications of ICT adoption during COVID-19. They posit privacy, ethics, trust, and security issues should be considered given the rapid use of technology by agencies and organizations for government services, employment, education, and healthcare.

References

1. American Journal of Managed Care [AJMC]. 2021, Jan 1. A Timeline of COVID-19 Developments in 2020. Accessed on September 9, 2021, at https://www.ajmc.com/view/a-timeline-of-covid19-developments-in-2020.
2. Bennett, DeeDee, Brenda Phillips, and Elizabeth Davis. (2017). "The Future of Accessibility in Disaster Conditions: How Wireless Technologies Will Transform the Life Cycle of Emergency Management." Futures Journal. 87(2017): 122–132.
3. Centers for Disease Control and Prevention [CDC]. 2021, Aug 9 Delta Variant: What We Know About the Science. CDC https://www.cdc.gov/coronavirus/2019-ncov/variants/delta-variant.html.
4. Chiauzzi, E., Clayton, A., & Huh-Yoo, J. (2020). Videoconferencing-Based Telemental Health: Important Questions for the COVID-19 Era from Clinical and Patient-Centered Perspectives. JMIR Mental Health, 7(12), e24021.
5. Cucinotta D, Vanelli M. WHO Declares COVID-19 a Pandemic. Acta Biomed. 2020 Mar 19;91(1):157–160. doi: https://doi.org/10.23750/abm.v91i1.9397. PMID: 32191675; PMCID: PMC7569573.
6. Digital Divide Council. 2019, February 22. What is the Digital Divide? Digital Divide Council posted by Carmen Steele. Access on September 9, 2021, at http://www.digitaldividecouncil.com/what-is-the-digital-divide/.
7. Donner, W. and H. Rodriguez. 2008. "Population Composition, Migration and Inequality: The Influence of Demographic Changes on Disaster Risk and Vulnerability." Social Forces 87(2):1089–1114.
8. Fatani, T. H. (2020). Student satisfaction with videoconferencing teaching quality during the COVID-19 pandemic. BMC Medical Education, 20(1), 1–8.
9. Fraustino, J. D., Liu, B. F., & Jin, Y. (2017). Social Media Use During Disasters 1: A Research Synthesis and Road Map. Social media and crisis communication, 283–295.
10. Griego, A. L., Flores, A. B., Collins, T. W., & Grineski, S. E. (2020). Social vulnerability, disaster assistance, and recovery: A population-based study of Hurricane Harvey in Greater Houston, Texas. International Journal of Disaster Risk Reduction, 51, 101766.
11. Ghafurian, M., Ellard, C., & Dautenhahn, K. (2021, August). Social companion robots to reduce isolation: A perception change due to COVID-19. In IFIP Conference on Human-Computer Interaction (pp. 43–63). Springer, Cham.
12. Hamideh, S., & Rongerude, J. (2018). Social vulnerability and participation in disaster recovery decisions: public housing in Galveston after Hurricane Ike. Natural Hazards, 93(3), 1629–1648.
13. Holshue, M. L., DeBolt, C., Lindquist, S., Lofy, K. H., Wiesman, J., Bruce, H., ... & Pillai, S. K. (2020). First case of 2019 novel coronavirus in the United States. New England Journal of Medicine.

14. Joshi, S., Collins, S., Kamino, W., Gomez, R., & Šabanović, S. (2020, November). Social robots for socio-physical distancing. In International Conference on Social Robotics (pp. 440–452). Springer, Cham.
15. Kaniasty, K. (2020). Social support, interpersonal, and community dynamics following disasters caused by natural hazards. Current opinion in psychology, 32, 105–109.
16. Lai, J., & Widmar, N. O. (2021). Revisiting the digital divide in the COVID-19 era. Applied Economic Perspectives and Policy, 43(1), 458–464.
17. Maragakis, L. L. (2020) Coronavirus, Social and Physical Distancing and Self-Quarantine. Johns Hopkins Medicine: Health accessed on August 12, 2021, at https://www.hopkinsmedic ine.org/health/conditions-and-diseases/coronavirus/coronavirus-social-distancing-and-self-qua rantine.
18. Mavhura, E., Manyena, B., & Collins, A. E. (2017). An approach for measuring social vulnerability in context: The case of flood hazards in Muzarabani district, Zimbabwe. Geoforum, 86, 103–117.
19. Morganstein, J. C., & Ursano, R. J. (2020). Ecological disasters and mental health: causes, consequences, and interventions. Frontiers in psychiatry, 11, 1.
20. Morrow, B. H. (1999). Identifying and mapping community vulnerability. Disasters, 23(1), 1–18.
21. Odekerken-Schröder, G., Mele, C., Russo-Spena, T., Mahr, D., & Ruggiero, A. (2020). Mitigating loneliness with companion robots in the COVID-19 pandemic and beyond: an integrative framework and research agenda. Journal of Service Management.
22. Okereafor, K., & Manny, P. (2020). Understanding cybersecurity challenges of telecommuting and video conferencing applications in the COVID-19 pandemic. Journal Homepage: http://ijmr.net.in, 8(6).
23. Peacock W. G., Van Zandt, S., Zhang, Y., &Highfield, W.E. (2014). Inequities in long-term housing recovery after disasters. Journal of the American Planning Association. 80(4), 356–371.
24. Puddister, K., & Small, T. A. (2020). Trial by Zoom? The Response to COVID-19 by Canada's Courts. Canadian Journal of Political Science/Revue canadienne de science politique, 53(2), 373–377.
25. Recht H. and Weber, L. (2021, January 17). Black Americans Are Getting Vaccinated at Lower Rates Than White Americans. Kaiser Health News Report. Accessed on September 9, 2021, at https://khn.org/news/article/black-americans-are-getting-vaccinated-at-lower-rates-than-white-americans/.
26. Roth Tran, B., & Sheldon, T. L. (2017). Same storm, different disasters: Consumer credit access, income inequality, and natural disaster recovery. Different Disasters: Consumer Credit Access, Income Inequality, and Natural Disaster Recovery (December 15, 2017).
27. Settembrino, M. R. (2017). " Sometimes You Can't Even Sleep at Night:" Social Vulnerability to Disasters among Men Experiencing Homelessness in Central Florida. International Journal of Mass Emergencies & Disasters, 35(2).
28. Tate, E., Rahman, M. A., Emrich, C. T., & Sampson, C. C. (2021). Flood exposure and social vulnerability in the United States. Natural Hazards, 106(1), 435–457.
29. Tavakoli, M., Carriere, J., & Torabi, A. (2020). Robotics, smart wearable technologies, and autonomous intelligent systems for healthcare during the COVID-19 pandemic: An analysis of the state of the art and future vision. Advanced Intelligent Systems, 2(7), 2000071.
30. Toeroek, I. (2017). Assessment of social vulnerability to natural hazards in Romania. Carpathian Journal of Earth and Environmental Sciences, 12(2), 549–562.
31. Ton, K. T., Gaillard, J. C., Adamson, C., Akgungor, C., & Ho, H. T. (2020). An empirical exploration of the capabilities of people with disabilities in coping with disasters. International Journal of Disaster Risk Science, 11(5), 602–614.

32. Viswanathan, R., Myers, M. F., & Fanous, A. H. (2020). Support groups and individual mental health care via video conferencing for frontline clinicians during the COVID-19 pandemic. Psychosomatics, 61(5), 538–543.
33. World Health Organization [WHO]. (2020a, Dec 31). WHO issues its first emergency use validation for a COVID-19 vaccine and emphasizes need for equitable global access. World Health Organization. Accessed on September 9, 2021, at https://www.who.int/news/item/31-12-2020-who-issues-its-first-emergency-use-validation-for-a-covid-19-vaccine-and-emphas izes-need-for-equitable-global-access.
34. World Health Organization [WHO]. (2020b, July 17). A Guide to WHO's Guidance on COVID-19. World Health Organization. Accessed on September 9, 2021, at https://www.who.int/news-room/feature-stories/detail/a-guide-to-who-s-guidance.
35. World Health Organization [WHO]. (2020c, April 4). Coronavirus disease 2019 (COVID-19): situation report, 75. World Health Organization. Accessed on September 9, 2021, at https://www.who.int/publications/m/item/situation-report---75.
36. World Health Organization. (2020d, March 7). Critical preparedness, readiness, and response actions for COVID-19: interim guidance. World Health Organization. Accessed on September 9, 2021, at https://apps.who.int/iris/handle/10665/331422.
37. World Health Organization [WHO]. (2020e, Jan 29). WHO confirms first cases of novel coronavirus (2019-nCoV) in the Eastern Mediterranean Region. World Health Organization Regional Office for the Eastern Mediterranean Accessed on September 9, 2021, at http://www.emro.who.int/media/news/who-confirms-first-cases-of-novel-coronavirus-2019-ncov-in-the-eastern-mediterranean-region.html.
38. World Health Organization [WHO]. (2020f, Jan 25). 2019-nCoV outbreak: first cases confirmed in Europe. World Health Organization Regional Office for Europe Accessed on September 9, 2021, at https://www.euro.who.int/en/health-topics/health-emergencies/pages/news/news/2020/01/2019-ncov-outbreak-first-cases-confirmed-in-europe.
39. World Health Organization [WHO]. 2021. WHO Coronavirus (COVID-19) Dashboard. Accessed on September 9, 2021, at https://covid19.who.int.

DeeDee Bennett Gayle, Ph.D. is an Associate Professor in the College of Emergency Preparedness, Homeland Security, and Cybersecurity at the University at Albany, State University of New York. She broadly examines the influence and integration of advanced technologies on the practice of emergency management, and for use by vulnerable populations.

She has secured research grants and contracts, including from the National Science Foundation, Federal Emergency Management Agency, and the Department of Homeland Security. Her work is published in various journals, and she has presented at several conferences related to emergency management, disability, wireless technology, and future studies.

Dr. Bennett Gayle received her Ph.D. from Oklahoma State University in Fire and Emergency Management. She has a unique academic background having received both her M.S. in Public Policy and B.S. in Electrical Engineering from the Georgia Institute of Technology. She is an Advisory Board Member for the Institute for Diversity and Inclusion in Emergency Management (I-DIEM), a member of the Social Science Extreme Events Reconnaissance (SSEER) and Interdisciplinary Science Extreme Events Reconnaissance (ISEER), within the NSF-FUNDED CONVERGE initiative.

Xiaojun Yuan, Ph.D., is an Associate Professor in the College of Emergency Preparedness, Homeland Security, and Cybersecurity at the University at Albany, State University of New York. Her research interests include both Human Computer Interaction and Information Retrieval, with the focus on user interface design and evaluation and human information behavior.

She has received various grants and contracts, including from the Institute of Museum and Library Servcices, SUNY seed grant, Initiatives For Women Program at University at Albany, and New York State Education Department.

She published extensively in journals in information retrieval and human computer interaction (JASIS&T, IP&M, Journal of Documentation, etc.), and conferences in computer science and information science (ACM SIGIR, ACM SIGCHI, ACM CHIIR, ASIS&T, etc.).

Dr. Yuan received her Ph.D. from Rutgers University at the School of Communication and Information, and Ph.D. from Chinese Academy of Sciences in the Institute of Computing Technology. She received her M.S. in Statistics from Rutgers University and M.E. and B.E. in Computer Application from Xi'an University of Science & Technology in China. She serves as an Editorial Board Member of Aslib Journal of Information Management (AJIM), and a Board Member of the International Chinese Association of Human Computer Interaction. She is a member of the Association for Information Science and Technology (ASIS&T), the Association for Computing Machinery (ACM) and the Institute of Electrical and Electronics Engineers (IEEE).

Technological Innovations in Response to COVID-19: Research Agenda Considering Marginalized Populations

DeeDee Bennett Gayle, Xiaojun Yuan, Elisabeth Dubois, and Thora Knight

Abstract

This chapter presents an interdisciplinary research agenda for understanding the impacts COVID-19 response has had on our use of technology. The widespread unprecedented mandates on social distancing have forced a large majority of nearly 330 million Americans to rely on technology for work, education, and crucial societal functions. Using the Ecology framework, this research agenda identifies the domains of influence for the use of technology—from the individual and community to the organizational and societal levels. This chapter proposes a series of questions focused on the framework and offers a catalog of research questions as a launchpad for future research. This agenda serves as a guide for scholars and practitioners interested in understanding the influence of technology on the expansion or reduction of vulnerabilities for socially marginalized populations. The findings of the review suggest an increase in research on meso-, exo-, techno-, and macro-level interventions of technology use during COVID-19 and that some marginalized populations are not researched as much as others.

Keywords

Technology • COVID-19 • Social vulnerability • Research agenda

D. Bennett Gayle (✉) · X. Yuan · E. Dubois · T. Knight
University at Albany, State University of New York, Albany, NY, USA
e-mail: dmbennett@albany.edu

2.1   Introduction

The COVID-19 pandemic was unprecedented in several ways, the sheer number of people contracting the virus, the disproportionate impacts on older adults, people with pre-existing conditions, and the global reach. While the direct effect of contracting the virus was often the first topic of research studied, the pandemic had other indirect effects, namely the reliance on technology for continuity of daily life. In this chapter, a dedicated research agenda is proposed focused on the challenges and successes in the use, accessibility, and acceptance of technology during the pandemic by marginalized populations.

Researchers have previously proposed a general research agenda on Technological Innovations in Response to COVID-19.[1] This chapter builds on that work and narrowly focuses on socially vulnerable populations. During disasters, the most marginalized in society often face more difficulties during preparedness, response, and recovery [31, 37, 51, 80]. Evidence of their disparate burden has been documented following the impacts from natural hazards, human-induced disaster events, and in response to infectious diseases [46, 66, 81, 120]. *Socially vulnerable populations* is a term used to broadly classify often marginalized individuals and groups that may be more prone to the disproportionate effects of disasters due to an inability to respond to, cope with or anticipate these events [121]. These populations can be found in almost all regions or countries in the world, however, the individuals (or groups) included often differ based on the social construction of the society [38]. It is important to note that one's membership in one (or more) of the identified socially vulnerable populations does not guarantee that they will be more vulnerable than the general population during a disaster [38, 121]. In fact, the identified groups, as previously stated are grouped under this term for broad classification of a reoccurring trend. These groups may include children [92, 93], older adults [22, 75], historically marginalized racial and ethnic minorities [18, 36], people with disabilities [11, 13, 14], low-income communities [37], and gendered minorities [51], among others [38]. During the pandemic, researchers examined the vulnerability of these groups in terms of exposure to the virus, increased risk of being furloughed due to social distancing, mental health concerns, and regarding challenges in use, accessibility, or acceptance and reliance of technology.

Technologies were used in various ways during this pandemic, including to streamline government and organizational services, to assist with lifesaving measures in the hospital setting, and to provide continuity of educational opportunities [15, 30]. However, this research agenda is focused on the use of information and communications technologies

---

[1] Bennett et al. [12]. NSF CONVERGE Working Group, COVID-19 Global Research Registry for Public Health and Social Sciences Technological Innovations in Response to COVID-19. This COVID-19 Working Group effort was supported by the National Science Foundation-funded Social Science Extreme Events Research (SSEER) Network and the CONVERGE facility at the Natural Hazards Center at the University of Colorado Boulder (NSF Award #1,841,338).

(ICT) used at the household or individual level by the consumer for their personal use. For example, ICT used at the household level to monitor individuals would not be included in this study. These technologies may include communication devices, web conferencing systems, personal contact tracing devices, telehealth/ e-health systems, specialized equipment to enable an individual to perform job or academic tasks, or technology to assist an individual in daily living activities.

Significant literature suggests that socially vulnerable populations may face more barriers to technology adoption, prior to and during the pandemic [15, 39, 67]. Namely, researchers discuss the potential for a *digital divide* or gap within the population based on access to different technologies [39, 67, 100]. This gap exists between those who are marginalized, including low income, older adults, people with disabilities, those who lack access to power (such as children), and those for which discriminatory practices present a barrier (such as racial and ethnic minorities) [100, 117, 124]. The research on the digital divide has presented findings for the individual, organizations [45, 60], government and policy [32], as well as cultural shifts [100].

This chapter will begin with an exploratory literature review to identify the types of research that have occurred since 2020 which studies at least one of a select subset of socially vulnerable populations and their use of technology during COVID-19. The populations examined in this review are children, older adults, people with disabilities, racial and ethnic minorities (in a country or region), low-income, gender, or general marginalized populations. The studies will then be classified in terms of the environmental systems in which individuals interact using the ecological systems theory.

2.2 Ecological Systems Theory

Initially used to explain human development, the ecological systems theory developed by Bronfenbrenner (and since adapted), suggests that development occurs as a result of the direct interaction between the individual and their surroundings [19]. The model is used here to explore the changes that occur because of the interactions between technology and five environments given the near-ubiquitous global use of technology in response to COVID-19. The five environments include the Micro-system, Mesosystem, Exo-system, Techno-subsystem, and Macrosystem (Fig. 2.1).

The Ecological Systems Theory was also used by the general research agenda on Technological Innovations in Response to COVID-19 [12]. Each of the five environments was defined as follows:

- *Micro-level*: Refers to the use of technology at the individual or household level for individual use.
- *Meso-level:* Consists of interconnections between the Micro and Exo-level. Such as technology used to connect with schools, hospital, or community/neighborhood.

Fig. 2.1 Image of the ecological systems theory

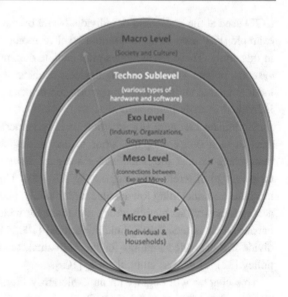

- *Exo-level:* Involves the use of technology for employment or to reach government or organizational services.
- *Techno-Sublevel:* Refers to differences in the use of technology for any purpose at the individual level that are caused because of the features of the technology itself.
- *Macrosystem:* Describes the overarching culture that influences the use of technology, in this study, specifically due to the use by a socially vulnerable population.

In this literature review, the contents of the articles will be segmented using the above-listed environments, as well as the dimension of vulnerability.

2.3   Literature Review

This systematic mapping review of the literature is used to identify current literature about the use of technology during the COVID-19 pandemic by socially vulnerable populations. The literature review revealed gaps in the research and mapped the current studies in terms of the five environments described in the Ecological Systems Theory and by the dimension of vulnerability discussed. This exploratory review is used as the basis for the subsequently proposed research agenda.

This literature review used the Web of Science database to unearth peer-reviewed journal articles with the following topics: COVID-19, Technology, and Socially Vulnerable Populations. To search through topics of articles on COVID-19 the following terms were used: ("COVID-19" OR "coronavirus disease 2019" OR "SARS-CoV-2"). Additionally, articles were searched by topic on the broad term: "technology." Furthermore, the articles were subject to a third search term by topic: ("children" OR "race" OR "socioeconomic" OR "elderly" OR "low income" OR "marginalized" OR "people with disabilities" OR "gender"). Given there are so many different subpopulations and groups that are considered socially vulnerable during disasters, the list was narrowed to include children, older adults, racial and ethnic minorities, low-income populations, people with disabilities, and gender. In this article the term 'marginalized populations' is used interchangeably with socially vulnerable populations, even though it is acknowledged that all marginalized populations may not be socially vulnerable to disasters. Thus, the term 'marginalized' was included as a search term. The search was refined to include only articles published in 2020 or 2021 and written in English. A total of 558 articles were identified.

During the first-round review of abstracts and titles, the articles were selected for inclusion based on the research focus. A total of 284 articles were included. During the second-round review, the articles were reviewed in their entirety and articles not fitting the criteria were excluded. A total of 188 articles were excluded from the review using four coders. The reasons for exclusion included: articles not found/ not retrieved (n = 12); not technology-focused (n = 64); not about socially vulnerable populations (n = 50); review articles/ commentary/no outcomes (n = 28); comparison between low- and middle-income countries (n = 6); not focused on technology that will be used at the individual level (e.g., x rays) (n = 22); not about COVID-19 pandemic (n = 3); about the organization use of technology (n = 3). The final list of articles was mapped by the methodology, type of technology studied, Ecological Systems Theory Level, dimension of social vulnerability included (Fig. 2.2).

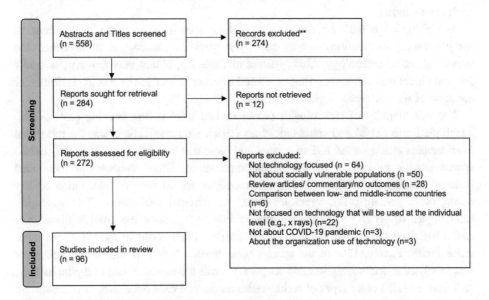

Fig. 2.2 Overview of the screening process for the review

Finally, 96 articles were selected for inclusion. The articles were segmented by the dimension of social vulnerability included, where no more than 36 articles were found in any one category.

2.4   Results

The segmented articles included 37.5% children, 28% older adults, 17.7% people with disabilities, 7.2% race or ethnicity, 20.8% gender, 8.3% low-income households or communities, and 3.1% general marginalized populations. General marginalized populations was the term used to describe papers that only broadly used the term marginalized populations in the paper without narrowing in on a specific subpopulation(s). Nearly 22% of the articles covered multiple dimensions of vulnerability, such as children with disabilities or older adults and gender. The category for gender represented both studies about women or men specifically and studies about the LGBT community. The articles were also identified by the ecological framework level represented in the research. Approximately 74% of the articles focused on micro-level interventions for increased or improved technology use by at least one socially vulnerable population. Nearly 15% were focused on the meso-level, 2% at the exo-level, 4% were focused on differences based on the type of technology used and one study focused on societal level or cultural changes needed for increased or improved use. Table 2.1 shows how the articles were segmented. While each article was only categorized by the framework level once, some articles reflected research that covered one or more subpopulations (e.g. children and people with disabilities, or older adults and gender).

As highlighted in Table 2.1, the majority of the articles focus on micro-level (individual and household) interventions such as training on specific technology or surveys about the individual use of technology. Also apparent in Table 2.1, is that very few articles solely focused on cultural shifts occurring or needed to occur to enable the use, accessibility, or adoption of specific technologies.

The vast majority of technologies (46%) studied were online learning (25%), Telehealth platforms (21%) and communications (19%). Communications was the term used to encompass articles about ICT in general and those that mentioned social media, mobile communication devices or mobile device applications. Other technologies mentioned included reality-based: either virtual, augmented, or mixed reality (5%), video conferencing (4%), gaming (4%), wearables (3%), and artificial intelligence (3%). Emerging technologies or general technologies were discussed in 3% of the articles, these were coded based on the use of the term in the article, which could include ICT and other technologies. Finally, 13% of the articles were labeled 'other' if they mentioned other factors related to technology use and adoption such as the digital divide or digital literacy, or if they included other types of technologies mentioned only once such as e-mentoring, memory cades, virtual volunteering, or online media. Figure 2.3 shows a pie chart of the

Table 2.1 Breakdown of articles review by ecological framework and dimension of vulnerability

	Children (n = 36)	Older adults (n = 27)	People with disabilities (n = 17)	Race or ethnicity (n = 7)	Gender (n = 20)	Low income (n = 8)	General marginalized populations (n = 3)
Micro-level (n = 71)	[2, 8, 9, 16, 20, 21, 27–29, 34, 35, 42, 44, 53, 54, 58, 64, 76, 84, 85, 94, 103, 107, 114, 12723, 41, 56– 91]	[5, 10, 24, 43, 47, 49, 71, 72, 79, 82, 89, 91, 98, 106, 109, 111, 112, 118, 69, 102, 123]	[2, 5, 16, 42, 52, 78, 82, 84, 113, 68,]	[6, 10, 72, 104, 119]	[1, 4, 40, 53, 61, 63, 73, 90, 104, 105, 110, 115, 116, 65,]	[40, 72, 74, 119]	[17, 101]
Meso-level (n = 14)	[70, 86, 97, 99]	[7, 50, 125]	[7, 97, 99]	[62, 87]	[26, 55, 95, 126]	[48, 62]	
Exo-level (n = 2)						[3]	[88]
Techno-level (n = 4)	[96]	[59, 77]			[33]		
Macro-level (n = 1)		[83]			[83]		
Unknown (n = 3)	[25, 122]		[25, 108, 122]				

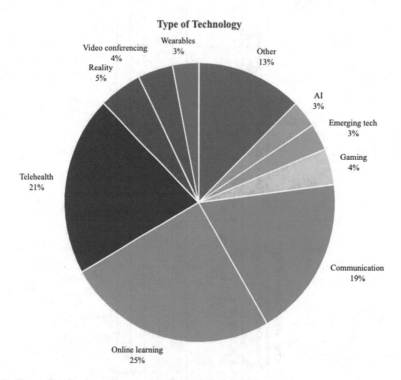

Fig. 2.3 Type of technology discussed in the included articles

type of technology discussed in each of the articles.

Of the articles that specifically mentioned the study location (n = 70), most of them were based in the United States (30%). Twenty-eight countries were identified as study locations for research on the use of technology. Two articles were about regional studies in Europe and the Gaza Strip, and there was one article about a comparative study between Australia and USA. Using the United Nations geoscheme to identify regions, the article locations were segmented as follows: 4% Africa, 34% Americas, 33% Asia, 27% Europe, 3% Oceania, including the comparison study.

The research articles also cut across several methods including 54.2% quantitative analysis primarily using a survey, 26% qualitative analysis of which most were interviews, and 13.5% mixed methods with a majority conducting an experimental (or quasi-experimental) design). Approximately 6.3% of articles were not categorized as quantitative, qualitative, or mixed methods because they were proposing a new method or new technology. Participant sizes ranged from 10 to over 60,000 between quantitative and qualitative studies, omitting the case studies, which typically analyzed less than ten.

2.5   Research Agenda

The COVID-19 pandemic social distancing guidelines and shifts to work, educate, and play from home have increased our need and reliance on ICT. The socially vulnerable populations of the world are often the most marginalized, where there are significant barriers to use, adoption, or accessibility with many ICTs. This proposed research agenda is based on the gaps in the literature as it relates to the use of technology by socially marginalized populations during the pandemic.

With the literature centered at the micro-level research a couple of questions emerge:

1. What laws, policies, or procedures implemented by the government changed how socially vulnerable populations use or have access to technology?
2. Were all information and communication technologies (ICTs) equally accessible for the socially vulnerable during the pandemic? Are some easier to use than others?

Given the research was primarily focused on children or older adults, there are several added research questions:

1. What, if any challenges were faced by racial and ethnic minority populations?
2. What considerations were made to aid low-income populations with use and access to necessary technology during the pandemic?
3. Did the reliance on technology empower people with disabilities? What were the differences based on an individual's ability?
4. What societal or cultural level challenges were barriers for marginalized populations in general? How did these differ according to country or region?
5. Given the number of articles focused on gender, what were the cultural or religious considerations that may have presented a barrier for use, accessibility, or adoption of technology based on gender?

Finally, due to the type of technologies found in this review, a few more questions emerge:

1. How did newer technologies, such as virtual reality or gaming benefit or challenge socially vulnerable populations during the pandemic?
2. Given the significant use of social media in general, in what unique ways were these platforms used by socially vulnerable populations during the pandemic?
3. Due to the increased use of artificial intelligence within many systems, how were these technologies useful (or not) to increase adoption by marginalized populations?

2.6   Discussion

The number of articles found in this review confirms earlier research identifying socially vulnerable populations as having disproportionate impacts during disasters [31, 37, 51, 80]. In particular, the COVID-19 pandemic moved the world to social distance, often by staying home and employing online tools and technology for work, education, and other aspects of daily living.

From the review, one could argue that the collective move online primarily impacted the education sector and children, healthcare, and older adults (and others socially isolated). However, given literature on the digital divide, there appears to be a gap in the literature with regards to general marginalized populations, low-income households and racial and ethnic minorities [117, 124]. Furthermore, while most research focused on the user and how user characteristics might be improved to eliminate barriers to technology, there was a dearth of studies related to government or policy intervention, changes to the technology itself, or cultural shifts that may have been needed to enable technology use, accessibility, and adoption by all [32, 100].

While articles on gender were not lacking, the studies primarily focused on the needs of women. Only one article mentioned issues or concerns of the LGBT (lesbian, gay, bisexual and transgender) community. Similarly, articles on people with disabilities are lacking concerning the wide range of disabilities individuals may have. Many of these articles were general in consideration for all people with disabilities, which is not a homogenous community.

This study has limitations. The marginalized populations reviewed were not a comprehensive list and do not represent all marginalized populations globally. There is the potential for selection bias, whereby a researcher unintentionally selects articles that support their beliefs. To reduce the potential for this bias, four coders were used. Additionally, only one database was reviewed. This database was chosen because it provided the most significant contributions in the initial search, and contained a wide range of interdisciplinary journals, representing a good portion of the research available.

2.7   Conclusion

The general research agenda presented is not meant to be a comprehensive agenda, but a start to raising awareness about the gaps in the literature about technology use, accessibility, and adoption among marginalized populations worldwide. The findings suggest an increase in research on meso-, techno-, exo-, and macro-level interventions of technology use during COVID-19 using the ecology framework. The findings also suggest more research is needed focusing on the similarities in benefits or challenges among marginalized populations or on the specific dimensions of vulnerability such as race/ethnicity, low-income populations, or gender. Finally, a wider range of technologies should be investigated, beyond telehealth devices, tools, or platforms and tools for online learning.

Acknowledgements NHC CONVERGE Working Group, COVID-19 Global Research Registry for Public Health and Social Sciences Technological Innovations in Response to COVID-19. This COVID-19 Working Group effort was supported by the National Science Foundation-funded Social Science Extreme Events Research (SSEER) Network and the CONVERGE facility at the Natural Hazards Center at the University of Colorado Boulder (NSF Award #1841338).

References

1. Abid, T., Zahid, G., Shahid, N., & Bukhari, M. (2021). Online Teaching Experience during the COVID-19 in Pakistan: Pedagogy–Technology Balance and Student Engagement. *Fudan Journal of the Humanities and Social Sciences*, 1–25.
2. Abuzaid, S. (2021). Attitudes of intellectual disabled children's teachers towards E-learning during Corona pandemic. *Amazonia Investiga, 10*(40), 29–36.
3. Adarkwah, M. A. (2021). "I'm not against online teaching, but what about us?": ICT in Ghana post Covid-19. *Education and Information Technologies, 26*(2), 1665–1685.
4. Al Soub, T. F., Alsarayreh, R. S., & Amarin, N. Z. (2021). Students 'satisfaction with Using E-Learning to Learn Chemistry in Light of the COVID-19 Pandemic in Jordanian Universities. *International Journal of Instruction, 14*(3).
5. Ali, M. A., Alam, K., Taylor, B., & Ashraf, M. (2021). Examining the determinants of eHealth usage among elderly people with disability: The moderating role of behavioural aspects. *International Journal of Medical Informatics, 149*, 104411.
6. Anakwe, A., Majee, W., Noel-London, K., Zachary, I., & BeLue, R. (2021). Sink or Swim: Virtual Life Challenges among African American Families during COVID-19 Lockdown. *International Journal of Environmental Research and Public Health, 18*(8), 4290.
7. Arighi, A., Fumagalli, G. G., Carandini, T., Pietroboni, A. M., De Riz, M. A., Galimberti, D., & Scarpini, E. (2021). Facing the digital divide into a dementia clinic during COVID-19 pandemic: caregiver age matters. *Neurological Sciences, 42*(4), 1247–1251.
8. Arufe-Giráldez, V., Cachón Zagalaz, J., Zagalaz Sánchez, M., Sanmiguel-Rodríguez, A., & González Valero, G. (2020). Equipamiento y uso de Tecnologías de la Información y Comunicación (TIC) en los hogares españoles durante el periodo de confinamiento. Asociación con los hábitos sociales, estilo de vida y actividad física de los niños menores de 12 años. *Revista Latina de Comunicación Social, 78*, 183–204.
9. Asvial, M., Mayangsari, J., & Yudistriansyah, A. (2021). Behavioral intention of e-learning: A case study of distance learning at a junior high school in Indonesia due to the covid-19 pandemic. *Int. J. Technol, 12*, 54–64.
10. Bakshi, T., & Bhattacharyya, A. (2021). Socially Distanced or Socially Connected? Well-being through ICT Usage among the Indian Elderly during COVID-19. *Millennial Asia*, 0976399621989910.
11. Bennett, D. (2020). Five years later: Assessing the implementation of the four priorities of the Sendai framework for inclusion of people with disabilities. *International Journal of Disaster Risk Science, 11*(2), 155–166.

12. Bennett, D., Knight, T., Dubois, E., Khurana, P., Wild, D., Laforce, S., Yuan, X.-J. (2020). NSF CONVERGE Working Group, COVID-19 Global Research Registry for Public Health and Social Sciences: Technological Innovations in Response to COVID-19. This COVID-19 Working Group effort was supported by the National Science Foundation-funded Social Science Extreme Events Research (SSEER) Network and the CONVERGE facility at the Natural Hazards Center at the University of Colorado Boulder (NSF Award #1841338).

13. Bennett, D., LaForce, S., Touzet, C., & Chiodo, K. (2018). American Sign Language & Emergency Alerts: The Relationship between Language, Disability, and Accessible Emergency Messaging. *International Journal of Mass Emergencies & Disasters, 36*(1).

14. Bennett, D., Phillips, B. D., & Davis, E. (2017). The future of accessibility in disaster conditions: How wireless technologies will transform the life cycle of emergency management. *Futures, 87*, 122–132.

15. Bennett Gayle, D., X. Yuan, and T. Knight. (2021). Coronavirus Pandemic: The Use of Technology for Education, Employment, Livelihoods. *Journal Assistive Technology*. (accepted-forthcoming)

16. Bettini, E. A. (2020). COVID-19 pandemic restrictions and the use of technology for pediatric palliative care in the acute care setting. *Journal of Hospice & Palliative Nursing, 22*(6), 432–434.

17. Blom, A. G., Wenz, A., Cornesse, C., Rettig, T., Fikel, M., Friedel, S., ... & Krieger, U. (2021). Barriers to the large-scale adoption of the COVID-19 contact-tracing app in Germany: Survey study. *Journal of Medical Internet Research: JMIR, 23*(3), e23362.

18. Bolin, B., & Kurtz, L. C. (2018). Race, class, ethnicity, and disaster vulnerability. *Handbook of disaster research*, 181–203.

19. Bronfenbrenner, U. (1992). *Ecological systems theory*. Jessica Kingsley Publishers.

20. Çakıroğlu, S., Soylu, N., & Görmez, V. (2021). Re-evaluating the Digital Gaming Profiles of Children and Adolescents during the COVID-19 Pandemic: A Comparative Analysis Comprising 2 Years of Pre-Pandemic Data.

21. Cardona-Reyes, H., Muñoz-Arteaga, J., Villalba-Condori, K., & Barba-González, M. L. (2021). A Lean UX Process Model for Virtual Reality Environments Considering ADHD in Pupils at Elementary School in COVID-19 Contingency. *Sensors, 21*(11), 3787.

22. Cherry, K. E., Allen, P. D., Galea, S., & Dass-Brailsford, P. (2009). Older adults and natural disasters. *Crisis and disaster counseling: Lessons learned from Hurricane Katrina and other disasters*, 115–30.

23. Cockerham, D., Lin, L., Ndolo, S., & Schwartz, M. (2021). Voices of the students: Adolescent well-being and social interactions during the emergent shift to online learning environments. *Education and Information Technologies*, 1–19.

24. Cohen-Mansfield, J., Muff, A., Meschiany, G., & Lev-Ari, S. (2021). Adequacy of Web-Based Activities as a Substitute for In-Person Activities for Older Persons During the COVID-19 Pandemic: Survey Study. *Journal of medical Internet research, 23*(1), e25848.

25. Demers, M., Martinie, O., Winstein, C., & Robert, M. T. (2020). Active video games and low-cost virtual reality: an ideal therapeutic modality for children with physical disabilities during a global pandemic. *Frontiers in Neurology, 11*, 1737.

26. Diniz, C. S. G., Franzon, A. C. A., Fioretti-Foschi, B., Niy, D. Y., Pedrilio, L. S., Amaro Jr, E., & Sato, J. R. (2021). Communication Intervention Using Digital Technology to Facilitate Informed Choices at Childbirth in the Context of the COVID-19 Pandemic: Protocol for a Randomized Controlled Trial. *JMIR research protocols, 10*(5), e25016.

27. Dong, C., Cao, S., & Li, H. (2020). Young children's online learning during COVID-19 pandemic: Chinese parents' beliefs and attitudes. *Children and youth services review, 118*, 105440.

28. Drouin, M., McDaniel, B. T., Pater, J., & Toscos, T. (2020). How parents and their children used social media and technology at the beginning of the COVID-19 pandemic and associations with anxiety. *Cyberpsychology, Behavior, and Social Networking, 23*(11), 727–736.
29. Du, Y., Grace, T. D., Jagannath, K., & Salen-Tekinbas, K. (2021). Connected Play in Virtual Worlds: Communication and Control Mechanisms in Virtual Worlds for Children and Adolescents. *Multimodal Technologies and Interaction, 5*(5), 27.
30. Dubois, E., Bright, D., & Laforce, S. (2021). Educating Minoritized Students in the United States During COVID-19: How Technology Can be Both the Problem and the Solution. IT Professional, 23(2), 12–18.
31. Dubois, E. & Yuan, X. J. (2021). The mental state of Americans amid the COVID-19 crisis: How socially vulnerable populations face greater disparities during and after a crisis. Journal of Emergency Management, 19(9), 69–80.
32. Dzenowagis, J. (2019). Bridging the digital divide: linking health and ICT policy. In *Telehealth in the developing world (*pp. 9–26). CRC Press.
33. Elçi, A., & Abubakar, A. M. (2021). The configurational effects of task-technology fit, technology-induced engagement and motivation on learning performance during Covid-19 pandemic: An fsQCA approach. *Education and Information Technologies*, 1–19.
34. Escudero, P., Pino Escobar, G., Casey, C. G., & Sommer, K. (2021). Four-Year-Old's Online Versus Face-to-Face Word Learning via eBooks. *Frontiers in psychology, 12*, 450.
35. Farrugia, R. C., & Busuttil, L. (2021). Connections and disconnections between home and kindergarten: A case study of a 4-year old child's digital practices and experiences in early childhood. *British Journal of Educational Technology*.
36. Fothergill, A., Maestas, E. G., & Darlington, J. D. (1999). Race, ethnicity and disasters in the United States: A review of the literature. *Disasters, 23*(2), 156–173.
37. Fothergill, A., & Peek, L. A. (2004). Poverty and disasters in the United States: A review of recent sociological findings. *Natural hazards, 32*(1), 89–110.
38. Fordham, M., Lovekamp, W. E., Thomas, D. S., & Phillips, B. D. (2013). Understanding social vulnerability. *Social vulnerability to disasters, 2*, 1–29.
39. Fraustino, J. D., Liu, B. F., & Jin, Y. (2017). Social Media Use During Disasters 1: A Research Synthesis and Road Map. Social media and crisis communication, 283–295.
40. Gabr, H. M., Soliman, S. S., Allam, H. K., & Raouf, S. Y. A. (2021). Effects of remote virtual work environment during COVID-19 pandemic on technostress among Menoufia University Staff, Egypt: a cross-sectional study. *Environmental Science and Pollution Research*, 1–8.
41. Gaudreau, C., King, Y. A., Dore, R. A., Puttre, H., Nichols, D., Hirsh-Pasek, K., & Golinkoff, R. M. (2020). Preschoolers benefit equally from video chat, pseudo-contingent video, and live book reading: implications for storytime during the coronavirus pandemic and beyond. *Frontiers in Psychology, 11*, 2158.
42. Gefen, N., Steinhart, S., Beeri, M., & Weiss, P. L. (2021). Lessons Learned during a Naturalistic Study of Online Treatment for Pediatric Rehabilitation. *International Journal of Environmental Research and Public Health, 18*(12), 6659.
43. Giansanti, D., & Velcro, G. (2021, April). The Digital Divide in the Era of COVID-19: An Investigation into an Important Obstacle to the Access to the mHealth by the Citizen. In *Healthcare* (Vol. 9, No. 4, p. 371). Multidisciplinary Digital Publishing Institute.
44. Gerow, S., Radhakrishnan, S., S Akers, J., McGinnis, K., & Swensson, R. (2021). Telehealth parent coaching to improve daily living skills for children with ASD. *Journal of Applied Behavior Analysis, 54*(2), 566–581.
45. Ghislieri, C., Molino, M., & Cortese, C. G. (2018). Work and organizational psychology looks at the fourth industrial revolution: how to support workers and organizations? *Frontiers in psychology, 9*, 2365.

46. Godwin, E. E., Foster, V. A., & Keefe, E. P. (2013). Hurricane Katrina families: Social class and the family in trauma recovery. *The Family Journal, 21*(1), 15–27.

47. Gómez-Portes, C., Vallejo, D., Corregidor-Sánchez, A. I., Rodríguez-Hernández, M., Martín-Conty, J. L., Schez-Sobrino, S., & Polonio-López, B. (2021). A Platform Based on Personalized Exergames and Natural User Interfaces to Promote Remote Physical Activity and Improve Healthy Aging in Elderly People. *Sustainability, 13*(14), 7578.

48. González-Betancor, S. M., López-Puig, A. J., & Cardenal, M. E. (2021). Digital inequality at home. The school as compensatory agent. *Computers & Education, 168*, 104195.

49. Goodman-Casanova, J. M., Dura-Perez, E., Guzman-Parra, J., Cuesta-Vargas, A., & Mayoral-Cleries, F. (2020). Telehealth home support during COVID-19 confinement for community-dwelling older adults with mild cognitive impairment or mild dementia: survey study. *Journal of medical Internet research, 22*(5), e19434.

50. Hara, R., & Shimizu, T. (2021). The effect of room sound absorption on a teleconference system and the differences in subjective assessments between elderly and young people. *Applied Acoustics, 179*, 108050.

51. Hines, R. I. (2007). Natural disasters and gender inequalities: The 2004 tsunami and the case of India. *Race, Gender & Class*, 60–68.

52. Howard, P. L., & Sedgewick, F. (2021). 'Anything but the phone!': Communication mode preferences in the autism community. *Autism*, 13623613211014995.

53. Hu, X., Chiu, M. M., Leung, W. M. V., & Yelland, N. (2021). Technology integration for young children during COVID-19: Towards future online teaching. *British Journal of Educational Technology.*

54. Huertas-Abril, C. A. (2021). Developing Speaking with 21st Century Digital Tools in the English as a Foreign Language Classroom: New Literacies and Oral Skills in Primary Education. *Aula Abierta, 50*(2), 625–634.

55. Hsiao, Y. C. (2021). Impacts of course type and student gender on distance learning performance: A case study in Taiwan. *Education and Information Technologies*, 1–16.

56. Hwang, W. Y., & Hariyanti, U. (2020). Investigation of Students' and Parents' Perceptions of Authentic Contextual Learning at Home and Their Mutual Influence on Technological and Pedagogical Aspects of Learning under COVID-19. *Sustainability, 12*(23), 10074.

57. Iivari, N. (2020). Empowering children to make and shape our digital futures–from adults creating technologies to children transforming cultures. *The International Journal of Information and Learning Technology.*

58. Ionescu, C. A., Paschia, L., Gudanescu Nicolau, N. L., Stanescu, S. G., Neacsu Stancescu, V. M., Coman, M. D., & Uzlau, M. C. (2020). Sustainability analysis of the e-learning education system during pandemic period—covid-19 in Romania. *Sustainability, 12*(21), 9030.

59. Jaana, M., & Paré, G. (2020). Comparison of mobile health technology use for Self-Tracking between older adults and the general adult population in Canada: cross-sectional survey. *JMIR mHealth and uHealth, 8*(11), e24718.

60. Kamel, C., & Olausson, M. (2020). Value co-creation within the digital divide: how organizations can co-create value to maintain and attract older adults as their customers.

61. Kapser, S., Abdelrahman, M., & Bernecker, T. (2021). Autonomous delivery vehicles to fight the spread of Covid-19–How do men and women differ in their acceptance? *Transportation Research Part A: Policy and Practice, 148*, 183–198.

62. Kim, C. J. H., & Padilla, A. M. (2020). Technology for educational purposes among low-income Latino children living in a mobile park in Silicon Valley: A case study before and during COVID-19. *Hispanic Journal of Behavioral Sciences, 42*(4), 497–514.

63. Kim, S., Rosenblith, S., Chang, Y., & Pollack, S. (2020). Will ICMT Access and Use Support URM Students' Online Learning in the (Post) COVID-19 Era? *Sustainability, 12*(20), 8433.

64. Kouis, P., Michanikou, A., Anagnostopoulou, P., Galanakis, E., Michaelidou, E., Dimitriou, H., … & Yiallouros, P. K. (2021). Use of wearable sensors to assess compliance of asthmatic children in response to lockdown measures for the COVID-19 epidemic. *Scientific reports*, *11*(1), 1–11.

65. Korlat, S., Kollmayer, M., Holzer, J., Lüftenegger, M., Pelikan, E. R., Schober, B., & Spiel, C. (2021). Gender Differences in Digital Learning During COVID-19: Competence Beliefs, Intrinsic Value, Learning Engagement, and Perceived Teacher Support. *Frontiers in Psychology*, *12*, 849.

66. Lachlan, K. A., Burke, J., Spence, P. R., & Griffin, D. (2009). Risk perceptions, race, and Hurricane Katrina. *The howard journal of communications*, *20*(3), 295–309.

67. Lai, J., & Widmar, N. O. (2021). Revisiting the digital divide in the COVID-19 era. Applied Economic Perspectives and Policy, 43(1), 458–464.

68. Landi, D., Ponzano, M., Nicoletti, C. G., Cola, G., Cecchi, G., Grimaldi, A., … & Marfia, G. A. (2021). Patient's point of view on the use of telemedicine in multiple sclerosis: a web-based survey. *Neurological Sciences*, 1–9.

69. Lee, C. J., & Hsu, Y. (2021). Promoting the Quality of Life of Elderly during the COVID-19 Pandemic. *International Journal of Environmental Research and Public Health*, *18*(13), 6813.

70. Li, L., Flynn, K. S., DeRosier, M. E., Weiser, G., & Austin-King, K. (2021, June). Social-Emotional Learning Amidst COVID-19 School Closures: Positive Findings from an Efficacy Study of Adventures Aboard the SS GRIN Program. In *Frontiers in Education* (Vol. 6, p. 213). Frontiers.

71. Llorente-Barroso, C., Kolotouchkina, O., & Mañas-Viniegra, L. (2021). The Enabling Role of ICT to Mitigate the Negative Effects of Emotional and Social Loneliness of the Elderly during COVID-19 Pandemic. *International Journal of Environmental Research and Public Health*, *18*(8), 3923.

72. Lott, A., Sacks, H., Hutzler, L., Campbell, K. A., & Lajam, C. M. (2021). Telemedicine Utilization by Orthopedic Patients During COVID-19 Pandemic: Demographic and Socioeconomic Analysis. *Telemedicine and e-Health*.

73. Louail, B., Benarous, D., & Al-Enezi, H. (2020). Acceptance use Blackboard as a medium for E-learning during COVID-19 period for Saudi students: Case Northern Border University. *International Journal of Computer Science and Network Security*, 216–222.

74. Makgahlela, M., Mothiba, T. M., Mokwena, J. P., & Mphekgwana, P. (2021). Measures to Enhance Student Learning and Well-Being during the COVID-19 Pandemic: Perspectives of Students from a Historically Disadvantaged University. *Education Sciences*, *11*(5), 212.

75. Malik, S., Lee, D. C., Doran, K. M., Grudzen, C. R., Worthing, J., Portelli, I., … & Smith, S. W. (2018). Vulnerability of older adults in disasters: emergency department utilization by geriatric patients after Hurricane Sandy. *Disaster medicine and public health preparedness*, *12*(2), 184–193.

76. Marbán, J. M., Radwan, E., Radwan, A., & Radwan, W. (2021). Primary and Secondary Students' Usage of Digital Platforms for Mathematics Learning during the COVID-19 Outbreak: The Case of the Gaza Strip. *Mathematics*, *9*(2), 110.

77. Martínez-Alcalá, C. I., Rosales-Lagarde, A., Pérez-Pérez, Y. M., Lopez-Noguerola, J. S., Bautista-Díaz, M. L., & Agis-Juarez, R. A. (2021). The effects of Covid-19 on the Digital Literacy of the Elderly: Norms for Digital Inclusion. In *Frontiers in Education* (Vol. 6, p. 245). Frontiers.

78. Marotta, N., Demeco, A., Moggio, L., & Ammendolia, A. (2021). Why is telerehabilitation necessary? A pre-post COVID-19 comparative study of ICF activity and participation. *Journal of Enabling Technologies*.

79. Masoud, S. S., Meyer, K. N., Martin Sweet, L., Prado, P. J., & White, C. L. (2021). "We Don't Feel So Alone": A Qualitative Study of Virtual Memory Cafés to Support Social Connectedness Among Individuals Living with Dementia and Care Partners During COVID-19. *Frontiers in public health, 9*, 497.

80. Maurya, V. (2019). Natural Disasters, Psychological Well-Being and Resilience: Concerns related to Marginalized Groups. *International Journal of Research and Analytical Reviews*, 270–275.

81. Meltzer, G. Y., Avenbuan, O., Awada, C., Oyetade, O. B., Blackman, T., Erdei PhD, E., & Zelikoff PhD, J. T. (2020). Environmentally Marginalized Populations: the" perfect storm" for infectious disease pandemics, including COVID-19. *Journal of Health Disparities Research and Practice, 13*(4), 6.

82. McCausland, D., Luus, R., McCallion, P., Murphy, E., & McCarron, M. (2021). The impact of COVID-19 on the social inclusion of older adults with an intellectual disability during the first wave of the pandemic in Ireland. *Journal of Intellectual Disability Research.*

83. Mishra, M., & Majumdar, P. (2020). Social distancing during COVID-19: Will it change the Indian society? *Journal of Health Management, 22*(2), 224–235.

84. Mumbardó-Adam, C., Barnet-López, S., & Balboni, G. (2021). How have youth with Autism Spectrum Disorder managed quarantine derived from COVID-19 pandemic? An approach to families' perspectives. *Research in Developmental Disabilities, 110*, 103860.

85. Murphy, K. M., Cook, A. L., & Fallon, L. M. (2021). Mixed reality simulations for social-emotional learning. *Phi Delta Kappan, 102*(6), 30–37.

86. Obrero-Gaitán, E., Nieto-Escamez, F., Zagalaz-Anula, N., & Cortés-Pérez, I. (2021). An Innovative Approach for Online Neuroanatomy and Neuropathology Teaching Based on 3D Virtual Anatomical Models Using Leap Motion Controller During COVID-19 Pandemic. *Frontiers in Psychology, 12*, 1853.

87. Ojinnaka, C. O., & Adepoju, O. E. (2021). Racial and Ethnic Disparities in Health Information Technology Use and Associated Trends among Individuals Living with Chronic Diseases. *Population Health Management.*

88. Özdemir, V. (2021). Digital Is Political: Why We Need a Feminist Conceptual Lens on Determinants of Digital Health. *OMICS: A Journal of Integrative Biology, 25*(4), 249–254.

89. Padala, K. P., Wilson, K. B., Gauss, C. H., Stovall, J. D., & Padala, P. R. (2020). VA video connect for clinical care in older adults in a rural state during the COVID-19 pandemic: cross-sectional study. *Journal of Medical Internet Research, 22*(9), e21561.

90. Pal, D., & Patra, S. (2021). University students' perception of video-based learning in times of COVID-19: A TAM/TTF perspective. *International Journal of Human–Computer Interaction, 37*(10), 903–921.

91. Pan, S. L., Cui, M., & Qian, J. (2020). Information resource orchestration during the COVID-19 pandemic: A study of community lockdowns in China. *International Journal of Information Management, 54*, 102143.

92. Peek, L. (2008). Children and disasters: Understanding vulnerability, developing capacities, and promoting resilience—An introduction. *Children Youth and Environments, 18*(1), 1–29.

93. Peek, L., Abramson, D. M., Cox, R. S., Fothergill, A., & Tobin, J. (2018). Children and disasters. In *Handbook of disaster research* (pp. 243–262). Springer, Cham.

94. Popyk, A., & Pustułka, P. (2021). Transnational Communication between Children and Grandparents during the COVID-19 Lockdown. The Case of Migrant Children in Poland. *Journal of Family Communication*, 1–16.

95. Ramaswamy, A., Yu, M., Drangsholt, S., Ng, E., Culligan, P. J., Schlegel, P. N., & Hu, J. C. (2020). Patient satisfaction with telemedicine during the COVID-19 pandemic: retrospective cohort study. *Journal of medical Internet research, 22*(9), e20786.

96. Rochmah, N., Faizi, M., Hisbiyah, Y., Triastuti, I. W., Wicaksono, G., & Endaryanto, A. (2021). Quality of Life Differences in Pre-and Post-Educational Treatment in Type 1 Diabetes Mellitus During COVID-19. *Diabetes, Metabolic Syndrome and Obesity: Targets and Therapy, 14*, 2905.

97. Roitsch, J., Moore, R. L., & Horn, A. L. (2021). Lessons learned: What the COVID-19 global pandemic has taught us about teaching, technology, and students with autism spectrum disorder. *Journal of Enabling Technologies*.

98. Sacco, G., Lléonart, S., Simon, R., Noublanche, F., Annweiler, C., & TOVID Study Group. (2020). Communication technology preferences of hospitalized and institutionalized frail older adults during COVID-19 confinement: cross-sectional survey study. *JMIR mHealth and uHealth, 8*(9), e21845.

99. Semprino, M., Fasulo, L., Fortini, S., Molina, C. I. M., González, L., Ramos, P. A., ... & Caraballo, R. (2020). Telemedicine, drug-resistant epilepsy, and ketogenic dietary therapies: A patient survey of a pediatric remote-care program during the COVID-19 pandemic. *Epilepsy & Behavior, 112*, 107493

100. Scheerder, A., Van Deursen, A., & Van Dijk, J. (2017). Determinants of Internet skills, uses and outcomes. A systematic review of the second-and third-level digital divide. *Telematics and informatics, 34*(8), 1607–1624.

101. Schubel, L. C., Wesley, D. B., Booker, E., Lock, J., & Ratwani, R. M. (2021). Population subgroup differences in the use of a COVID-19 chatbot. *NPJ digital medicine, 4*(1), 1–3.

102. Schiaffini, R., Barbetti, F., Rapini, N., Inzaghi, E., Deodati, A., Patera, I. P., ... & Cianfarani, S. (2020). School and pre-school children with type 1 diabetes during Covid-19 quarantine: The synergic effect of parental care and technology. *diabetes research and clinical practice, 166*, 108302.

103. Schmidtberg, L. C., Grindle, C., Hersh, D. S., Rowe, C., Healy, J., & Hughes, C. D. (2021). Telehealth in Pediatric Surgical Subspecialties: Rapid Adoption in the Setting of COVID-19. *Telemedicine and e-Health*.

104. Seo, H., Altschwager, D., Choi, B. Y., Song, S., Britton, H., Ramaswamy, M., ... & Yenugu, L. (2021). Informal Technology Education for Women Transitioning from Incarceration. *ACM Transactions on Computing Education (TOCE), 21*(2), 1–16.

105. Seo, H., Britton, H., Ramaswamy, M., Altschwager, D., Blomberg, M., Aromona, S., ... & Wickliffe, J. (2020). Returning to the digital world: Digital technology use and privacy management of women transitioning from incarceration. *new media & society*, 1461444820966993.

106. Shao, D., & Lee, I. J. (2020). Acceptance and Influencing Factors of Social Virtual Reality in the Urban Elderly. *Sustainability, 12*(22), 9345.

107. Sharawat, I. K., & Panda, P. K. (2021). Caregiver satisfaction and effectiveness of teleconsultation in children and adolescents with migraine during the ongoing COVID-19 pandemic. *Journal of Child Neurology, 36*(4), 296–303.

108. Spencer, P., Van Haneghan, J. P., Baxter, A., Chanto-Wetter, A., & Perry, L. (2021). "It's ok, mom. I got it!": Exploring the experiences of young adults with intellectual disabilities in a postsecondary program affected by the COVID-19 pandemic from their perspective and their families' perspective. *Journal of Intellectual Disabilities*, 17446295211002346.

109. Sinha, S., Verma, A., & Tiwari, P. (2021). Technology: Saving and Enriching Life During COVID-19. *Frontiers in psychology, 12*, 698.

110. Sitar-Tăut, D. A. (2021). Mobile learning acceptance in social distancing during the COVID-19 outbreak: The mediation effect of hedonic motivation. *Human Behavior and Emerging Technologies, 3*(3), 366–378.

111. Sun, P. C., Morrow-Howell, N., Pawloski, E., & Helbach, A. (2021). Older Adults' Attitudes Toward Virtual Volunteering During the COVID-19 Pandemic. *Journal of Applied Gerontology*, 07334648211006978.

112. Sun, S., Folarin, A. A., Ranjan, Y., Rashid, Z., Conde, P., Stewart, C., ... & Haro, J. M. (2020). Late onset infectious complications and safety of tocilizumab in the management of COVID-19.

113. Szczepura, A., Holliday, N., Neville, C., Johnson, K., Khan, A. J. K., Oxford, S. W., & Nduka, C. (2020). Raising the digital profile of facial palsy: national surveys of patients' and clinicians' experiences of changing UK treatment pathways and views on the future role of digital technology. *Journal of medical Internet research*, *22*(10), e20406.

114. Tambyraja, S. R., Farquharson, K., & Coleman, J. (2021). Speech-Language Teletherapy Services for School-Aged Children in the United States During the COVID-19 Pandemic. *Journal of Education for Students Placed at Risk (JESPAR)*, 1–21.

115. Tzafilkou, K., Perifanou, M., & Economides, A. A. (2021). Development and validation of a students' remote learning attitude scale (RLAS) in higher education. *Education and Information Technologies*, 1–27

116. Upadhyay, U. D., & Lipkovich, H. (2020). Using online technologies to improve diversity and inclusion in cognitive interviews with young people. *BMC medical research methodology*, *20*(1), 1–10.

117. Wamuyu, P. K. (2017). Bridging the digital divide among low income urban communities. Leveraging use of Community Technology Centers. *Telematics and Informatics*, *34*(8), 1709–1720.

118. Wang, Z., Li, H., Wang, X., An, X., Li, G., Xue, L., ... & Gong, X. (2021). Situation of Integrated Eldercare Services with Medical Care in China. *Indian Journal of Pharmaceutical Sciences*, *83*(1), 140–152.

119. Wegermann, K., Wilder, J. M., Parish, A., Niedzwiecki, D., Gellad, Z. F., Muir, A. J., & Patel, Y. A. (2021). Racial and socioeconomic disparities in utilization of telehealth in patients with liver disease during COVID-19. *Digestive diseases and sciences*, 1–7.

120. Widener, P., & Gunter, V. J. (2007). Oil spill recovery in the media: Missing an Alaska Native perspective. *Society and Natural Resources*, *20*(9), 767–783.

121. Wisner, B., Blaikie, P., Cannon, T., & Davis, I. (2014). *At risk: Natural hazards, people's vulnerability and disasters*. Routledge.

122. Wong, P. W., Lam, Y., Lau, J. S., & Fok, H. (2020). The Resilience of Social Service Providers and Families of Children With Autism or Development Delays During the COVID-19 Pandemic—A Community Case Study in Hong Kong. *Frontiers in psychiatry*, *11*.

123. Wójcik, D., Szczechowiak, K., Konopka, P., Owczarek, M., Kuzia, A., Rydlewska-Liszkowska, I., & Pikala, M. (2021). Informal Dementia Caregivers: Current Technology Use and Acceptance of Technology in Care. *International journal of environmental research and public health*, *18*(6), 3167.

124. Yoon, H., Jang, Y., Vaughan, P. W., & Garcia, M. (2020). Older adults' Internet use for health information: Digital divide by race/ethnicity and socioeconomic status. *Journal of Applied Gerontology*, *39*(1), 105–110.

125. Zeghari, R., Guerchouche, R., Tran Duc, M., Bremond, F., Lemoine, M. P., Bultingaire, V., ... & König, A. (2021). Pilot Study to Assess the Feasibility of a Mobile Unit for Remote Cognitive Screening of Isolated Elderly in Rural Areas. *International Journal of Environmental Research and Public Health*, *18*(11), 6108.

126. Zhao, Y., Pinto Llorente, A. M., Sánchez Gómez, M. C., & Zhao, L. (2021). The Impact of Gender and Years of Teaching Experience on College Teachers' Digital Competence: An Empirical Study on Teachers in Gansu Agricultural University. *Sustainability*, *13*(8), 4163.

127. Zuo, M., Ma, Y., Hu, Y., & Luo, H. (2021). K-12 Students' Online Learning Experiences during COVID-19: Lessons from China. *Frontiers of Education in China, 16*(1), 1.

DeeDee Bennett Gayle, Ph.D. is an Associate Professor in the College of Emergency Preparedness, Homeland Security, and Cybersecurity at the University at Albany, State University of New York. She broadly examines the influence and integration of advanced technologies on the practice of emergency management, and for use by vulnerable populations.

She has secured research grants and contracts, including from the National Science Foundation, Federal Emergency Management Agency, and the Department of Homeland Security. Her work is published in various journals, and she has presented at several conferences related to emergency management, disability, wireless technology, and future studies.

Dr. Bennett Gayle received her Ph.D. from Oklahoma State University in Fire and Emergency Management. She has a unique academic background having received both her M.S. in Public Policy and B.S. in Electrical Engineering from the Georgia Institute of Technology. She is an Advisory Board Member for the Institute for Diversity and Inclusion in Emergency Management (I-DIEM), a member of the Social Science Extreme Events Reconnaissance (SSEER) and Interdisciplinary Science Extreme Events Reconnaissance (ISEER), within the NSF-FUNDED CONVERGE initiative.

Xiaojun Yuan, Ph.D., is an Associate Professor in the College of Emergency Preparedness, Homeland Security, and Cybersecurity at the University at Albany, State University of New York. Her research interests include both Human Computer Interaction and Information Retrieval, with the focus on user interface design and evaluation and human information behavior.

She has received various grants and contracts, including from the Institute of Museum and Library Servcices, SUNY seed grant, Initiatives For Women Program at University at Albany, and New York State Education Department.

She published extensively in journals in information retrieval and human computer interaction (JASIS&T, IP&M, Journal of Documentation, etc.), and conferences in computer science and information science (ACM SIGIR, ACM SIGCHI, ACM CHIIR, ASIS&T, etc.).

Dr. Yuan received her Ph.D. from Rutgers University at the School of Communication and Information, and Ph.D. from Chinese Academy of Sciences in the Institute of Computing Technology. She received her M.S. in Statistics from Rutgers University and M.E. and B.E. in Computer Application from Xi'an University of Science & Technology in China. She serves as an Editorial Board Member of Aslib Journal of Information Management (AJIM), and a Board Member of the International Chinese Association of Human Computer Interaction. She is a member of the Association for Information Science and Technology (ASIS&T), the Association for Computing Machinery (ACM) and the Institute of Electrical and Electronics Engineers (IEEE).

Elisabeth Dubois is a Ph.D. Candidate in the College of Emergency Preparedness, Homeland Security, and Cybersecurity. She completed a B.S. in Digital Forensics and an MBA from the University at Albany, in 2018 and 2020 respectively. Her research examines risk management across numerous disciplines including communication, public administration, business, IT, cybersecurity, and education. Within these areas, she is interested in risk management among socially vulnerable populations amid COVID-19 and operational cyber risk management in a global environment.

Elisabeth's passion for this area of research stems from her volunteer work and serving as the Director of Marketing at The Global Child, a school and safe-house for former street-working children in Cambodia. In this role, she has engaged with and taught former street children, learning the realities of living in a post-war era and has been driven to investigate ways to address the inequities across sectors, both academically and practically. Using her educational background and passion for community endeavors, she has co-founded The Global CyberTech Exchange to help educate and train underserved populations for careers in cybersecurity. Her unique background is grounded in both her educational and research pursuits, as well as putting what she has learned into practice.

Thora Knight is an information science doctoral candidate in the College of Emergency Preparedness, Homeland Security, and Cybersecurity (CEHC) at the University at Albany, State University of New York. She is a JD/MBA graduate from the State University of New York at Buffalo and holds a BS in business concentrating on information systems from the University of Phoenix. Her research broadly focuses on understanding digital technologies' privacy and security implications. Her current dissertation study examines the role of regulation in mitigating privacy concerns in the context of online behavioral advertising.

Adoption of Artificial Intelligence Technologies by Often Marginalized Populations

Xiaojun Yuan, DeeDee Bennett Gayle, Thora Knight, and Elisabeth Dubois

Abstract

Artificial Intelligence (AI) has found its application in many aspects of our lives. The COVID-19 pandemic has further allowed AI to play an increasingly important and beneficial role in our society, but it has also exposed the limitation of AI, particularly related to marginalized populations. This chapter first provides an overview of AI and equity pre-COVID, and then discusses what we know about AI during COVID-19. At the end, we conduct a systematic literature review to examine marginalized populations and their use of AI technologies during COVID-19. The populations examined in this review are children, older adults, people with disabilities, racial and ethnic minorities (in a country or region), low-income, gender, or general marginalized populations. The results indicate a huge gap for research on the use, adoption, and perception of AI technologies by communities that have previously experienced inequities in AI and COVID-19.

Keywords

Artificial intelligence • Marginalized populations • COVID-19 • Technology

X. Yuan (✉) · D. Bennett Gayle · T. Knight · E. Dubois
University at Albany, State University of New York, Albany, NY, USA
e-mail: xyuan@albany.edu

3.1 Introduction

The global COVID-19 pandemic has exposed many inequities for marginalized populations, with regards to technology [3]. During crises, disasters, and extreme events, marginalized populations are often more vulnerable and face more difficulty during preparedness, response, and recovery [13, 37]. These populations differ across countries and regions, but their vulnerability is correlated to marginalization. People with disabilities, low-income communities, children, older adults, women, and certain racial or ethnic minorities may be included among these groups [13].

For example, statistics show higher rates of infection, hospitalization, and death in various race and ethnic minority groups [12], but few research studies have focused on comprehensively understanding such inequalities and providing solutions to conquer COVID-19 and disparities in this regard. The modeling techniques and Artificial Intelligence (AI) solutions used during a crisis are used by the government and organizations to mitigate the crisis at hand [16, 32]. But these allowances are not always equitable, as some socially vulnerable populations do not have equal access to or use of AI, and the implementation or development of AI does not consider marginalized populations [21].

Although we lived in a digitally connected world prior to COVID-19, during the pandemic the impacts of technology and AI are ever more present. AI "enables computers and other automated systems to perform tasks that have historically required human cognition and what we typically consider human decision-making abilities." [25, para. 1]. AI has changed daily routines, travel, health care, transportation, education, security, agriculture and more allowing the public to evolve human capabilities. COVID-19 has brought to light the benefits of AI solutions in decision-making [31], but has also emphasized the need to protect the most vulnerable [29].

This chapter provides an overview of AI and equity pre-COVID-19, then discusses what we know about AI during COVID-19. Finally, we conduct a systematic literature review to examine marginalized populations and their use of AI technologies during COVID-19. The populations examined in this review are children, older adults, people with disabilities, racial and ethnic minorities (in a country or region), low-income, gender, or general marginalized populations. The results, discussion, and conclusion follow.

3.1.1 AI & Equity

Emerging technologies can either reduce or exacerbate inequities in society [33]. The technologies may include built-in bias based on the data or may be used to perpetuate inequities based on how it is implemented. These inequities may manifest among various sectors, education, employment, healthcare, etc. This section highlights a few areas of concern, providing examples of various frameworks to discuss AI and equity.

The use of AI in healthcare has the potential to perpetuate preexisting inequities because of built-in biases in the computer algorithm [15, 33]. For example, in the United States, the use of a popular health risk prediction tool relies on the health care costs attributed to the patient to predict needs going forward, which is biased against Black patients who may not receive the same level of care due to outside burdens (work, elder care, childcare, transportation, etc.). Using this tool within an AI health system will only perpetuate this bias. Therefore, researchers have suggested four considerations as a starting point for discussions regarding the equitable use and implementation of AI in public health: (1) the digital divide, (2) Algorithmic bias, (3) plurality of values across systems, and (4) Fair decision-making procedures [33]. Due to the current concerns in the United States during COVID-19, including increased awareness of structural racism and the rush for biomedical discoveries based on databases with poor representations, [9] have proposed the Ethical AI, Health Equity, and Racial Justice Framework integrated across the Lifecycle of AI development, Fig. 3.1.

Similarly, the use of AI in education suggests similar potentials for inequities [30, 34]. Noting that while there are benefits to the use and implementation of AI, there are societal and ethical concerns, as well [30]. A United Nations report identifies that the implementation of AI may not be easily transferable from country to country. "When Machine Learning algorithms are trained on a certain data set (let's say with students from a Western European country), then the result might not be directly applicable to students from other parts of the world. The training data set might be biased towards a

Fig. 3.1 Ethical AI, Health Equity, and Racial Justice Framework integrated across the Lifecycle of AI development (*Source* Dankwa-Mullan et al. [9])

certain group and therefore might discriminate unfairly when used on a different group [30, p. 33]." Figure 3.2 shows the education, equity, and AI (EEAI) framework proposed by researchers [34].

Additionally, [17] suggest four lenses by which to consider the potential inequities in AI for use in education, based on the socio-technical system design: (1) Factors Inherent to the Overall Socio-Technical System Design, (2) Use of Data that Reflect Historical Inequities, (3) Factors Inherent to the Underlying Algorithms, and (4) Interplay between Automated and Human Decision-making (see Fig. 3.3).

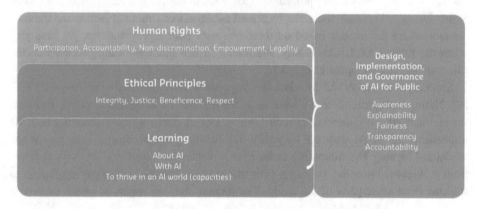

Fig. 3.2 Education, equity, and AI framework (*Source* Southgate [34])

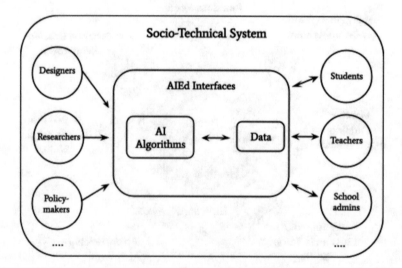

Fig. 3.3 Socio-technical system of AI for education (AIEd) (*Source* Holstein & Doroudi [17])

Finally, there are concerns about equity as AI is more readily used in government. One of the promising benefits of AI is in assistance with decision-making, thus its use in government for this purpose continues to grow. Researchers argue that new theoretical models are necessary to understand the impacts of such growth [14]. While new models are being developed, researchers present a chart articulating the level of human discretion needed as AI is incorporated into government. [38] proposed that low-level discretion is needed to automate tasks, medium-level discretion for decision-support or predictions, and high-level discretion is needed for the creation of new data. Among the researchers' concerns are unequal or unfair treatment of subpopulations, noting that "inequality in artificial discretion can be attributed both to the biases of the developers and implementers in the human system and to inequalities generated from historic human decisions (made possible through the exercise of discretion) that are embedded in the data used to train the system [38, p. 309]."

In addition to the consideration for potential inequities, research cautions about broader data and privacy concerns that may disproportionately impact the most marginalized populations. Within education, one of the primary challenges will be with "us[ing] personal data while ensuring that personally identifiable information and individual privacy preferences are protected [30, p. 33]." Concerning personal health information, certain data are not covered by health privacy regulations as customers unwittingly provide their data to third party markets such as wearable fitness devices, mobile phone devices, and ancestry data kits [36].

3.1.2 Equity in AI During Covid-19

At the forefront of using data is the accountability of AI in COVID-19 [4]. Yet, during a public health crisis, emergency legislation may trump privacy laws [26]. Those previously in positions of vulnerability due to financial state, race, age, or immigration status may have limited control over AI use and their own data protection, given the limited means to consent. The use and misappropriation of the personal data of the most vulnerable and impoverished persons in the world increased during the pandemic. For example, private sectors took the responsibility of designing interventions against COVID-19, which may lead to biases and inequalities due to the lack of transparency [35]. Those biases may come from the collection and measurement of data as well as the evaluation, aggregation, and deployment processes [21, 35], and might generate risks of racial discrimination within marginalized communities, emphasizing the importance of not only understanding the discrepancies on how AI is used by various populations, but privacy and data protection mechanisms expected or in place.

AI bias may be present during problem scoping, in the data used to train an AI system, or in the algorithm selection process themselves [21]. In a healthcare environment, the employment of AI methods may reproduce and amplify existing biases in medical data sets, by missing data from a minority group [5, 20]. As [27] pointed out, AI disequilibrium can be seen among vulnerable groups across three divides—the use divide, literacy gap, and bias problem. The use divide is the inequality resulting from a lack of infrastructure needed to boost connectivity, leading to increased challenges between those that can use AI and those that cannot [27]. The literacy gap refers to "language and digital literacy barriers" that may arise, whereupon many AI solutions are catered to specific populations or conversations. Similarly, this gap can cause people that are digitally illiterate and lack formal education to face job loss due to automation and AI. The bias problem is evident in training data, which widens existing inequalities placing populations at increased risk [27]. Although AI models are rigorous, the training data is often stand-in data that seeks to produce self-affirming results. These models can create feedback loops that manifest (or exacerbate) unequal outcomes for different groups of people [28]. For example, AI technologies able to diagnose Alzheimer's from auditory tests were found to contain bias given it only worked for English speakers of a particular Canadian dialect, causing long-term health impacts on non-white and/or English-speaking people [27].

During the pandemic, the reliance on technologies in society makes it critical to investigate how to develop advanced technologies and systems for socially vulnerable populations and engage them with such technologies and systems. AI proves to be able to offer necessary care and assessment to maintain health for older adults. As [2] reported, intelligent robots and smart wearable technologies contribute to reducing the spread of the COVID-19 virus by offering extended, accessible, and online delivery of assessment and rehabilitation services for older adults. However, most of the research or articles on vulnerability and AI investigate the impact in low-income and developing countries [27].

There is limited research regarding the use of AI, based on their gender, race, age, socioeconomic state, etc. This paper systematically reviews the current literature relevant to the use of artificial intelligence technologies by often marginalized populations. The review presented here is paramount given (1) the need to bridge the digital divide to ensure all groups are involved in the ICT space, (2) to curb the inequalities and oppression certain groups face in participating in the digital involved society, and (3) as a foundation to investigate how AI may help lessen vulnerability, while also identifying the gaps and how to solve the existing inequalities.

3.2   Methods

This PRISMA systematic review [24] is employed to identify current literature about the use of artificial intelligence during the COVID-19 pandemic by socially vulnerable populations.

This literature review used the Web of Science, ACM, Digital Library, and IEEE Xplore Digital Library databases to unearth peer-reviewed articles with the following topics: COVID-19, artificial intelligence, and marginalized populations. To search through topics of articles on COVID-19 the following terms were used: ("COVID-19" OR "coronavirus disease 2019" OR "SARS-CoV-2"). Additionally, articles were searched for by topic on the term: ("artificial intelligence" OR "intelligent"). Furthermore, the articles were subject to a third search term by topic: ("children" OR "race" OR "socioeconomic" OR "elderly" OR "low income" OR "marginalized" OR "people with disabilities" OR "gender"). Given there are a variety of different subpopulations and groups that are considered marginalized across countries, the list was narrowed to include children, elderly, socio-economic, racial and ethnic minorities, low-income populations, people with disabilities, and gender. The search was refined to include only peer-reviewed articles, written in English.

We performed three rounds of systematic selection in the selected databases. First, we searched predetermined keywords. Next, we screened the titles and abstracts using predetermined inclusion/exclusion criteria. Finally, we screened the full text of selected articles to ensure that they met the same inclusion/exclusion criteria.

Round 1: keyword search
On September 12, 2021, we searched the Topic field in the Web of Science database using the following three sets of keywords: ("artificial intelligence" OR "intelligent") AND ("children" OR "race" OR "socioeconomic" OR "elderly" OR "low income" OR "marginalized" OR "people with disabilities" OR "gender") AND ("COVID-19" OR "coronavirus disease 2019" OR "SARS-CoV-2"). This process produced a total of 90 results. On September 12, 2021, using the same sets of keywords, we searched in All metadata (including title, abstract and indexing terms) for all available years in the IEEE Xplore Digital Library, and found 64 results. We also searched in the titles and abstracts of the ACM Digital Library (from the ACM Full-text collection). This generated 3 results. Six duplicates were found. A total of 151 non-duplicate results remains.

Round 2: screening the titles and abstracts
Next, we screened the titles and abstracts of these 151 articles independently. This round of screening resulted in the removal of 101 articles from further analysis. This round of screening was based on the rationale that the focus of our systematic literature review was about socially vulnerable populations' use of artificial intelligence technologies. Other topics are outside of the scope of this review. Specifically, we removed an article if it met at least one of the following exclusion criteria:

- Use AI tools/techniques to analyze dataset or detect trends of infection, detect falls of older adults, and identify behavioral patterns ($n = 6$).
- Not empirical studies (e.g., literature review; book review; column/commentary/editorial; proceedings; $n = 27$).
- Not technology/AI or SVP related ($n = 10$).
- Not SVP related ($n = 28$).

- Not AI used by SVP (models, algorithms, frameworks) ($n = 30$).

After removing 101 articles, a total of 50 articles remains in the sample.

Round 3: screening the full text
We then screened the full text of all 50 remaining articles. During this round, we eliminated 43 more articles from our sample because they met at least one of the aforementioned exclusion criteria:

- Use AI tools/techniques to analyze dataset or detect trends of infection, detect falls of older adults, and identify behavioral patterns ($n = 5$).
- Not empirical studies (e.g., literature review; book review; column/commentary/editorial; $n = 5$).
- Not technology/AI or SVP related ($n = 1$).
- Not SVP related ($n = 11$).
- Not AI used by SVP (models, algorithms, frameworks) ($n = 18$).
- Not AI-related ($n = 3$).

A total of 7 articles remained in the final sample. Following the PRISMA guidelines for reporting systematic reviews, we summarize the selection process in Fig. 3.4. During the first-round review of abstracts and titles, the articles were selected for inclusion based on the research focus. A total of 50 articles were included. During the second-round review, the articles were reviewed in their entirety and mapped by the research question, methodology, type of technology studied, and the dimension of social vulnerability included. Finally, 7 articles were selected for inclusion.

3.3   Results

Our initial searches found 151 articles. Through multiple rounds of screening, we removed 144 of them from our final sample. The reasons for excluding these 144 articles are summarized in Table 3.1.

Our final sample included 7 articles (see Tables 3.2 and 3.3). Table 3.2 summarized the articles based on AI by population discussed, including children (3 articles), elderly (4 articles), people with disabilities (1 article).

Table 3.3 shows the coding framework and details of each article in the framework. These articles were published between 2020 and 2021. The year 2021 had 4 publications, suggesting increasing interest in this topic. The study aims of these articles fall into one of three major themes: (1) to develop a mobile app, systems, or robot; (2) to examine the relationship between dimensions of often marginalized populations and COVID-19

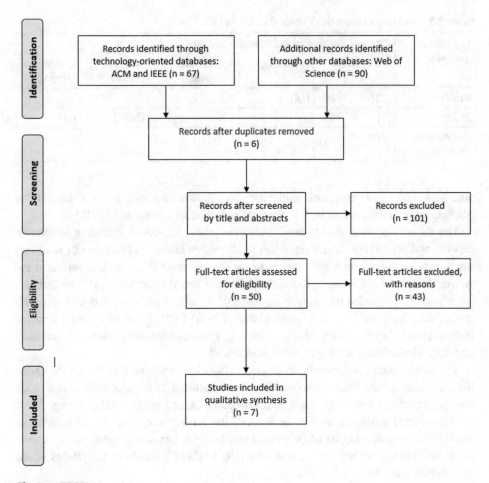

Fig. 3.4 PRISMA review

Table 3.1 Summary of the reasons for excluded articles

Reason for exclusion	n
Use AI tools/techniques to analyze datasets or detect trends of infection, detect falls of older adults, and identify behavioral patterns	11
Not empirical studies (e.g., literature review; book review; column/commentary/editorial	32
Not technology/AI or SVP related	11
Not SVP related	39
Not AI used by SVP (models, algorithms, frameworks)	48
Not AI related	3
TOTAL	144

Table 3.2 Summary of the articles based on SVP and AI

Population discussed	Artificial intelligence technologies included				
	Social robot	E-learning	Speech/image recognition	Wearable devices	IoT/deep learning
Children	[22]	[18]	[1]		
Elderly	[7]			[8, 19]	[6]
People with disabilities	[22]				

outcomes; and (3) to recognize difficulties and factors that may affect E-learning or perceptions of smart devices by the marginalized populations during COVID-19.

The majority of these studies used qualitative research methods, including interviews, surveys, and observation. The sample size of all studies in our final sample was relatively small, ranging from 5 to 4109, with most on the low end. Three studies described the system designed for the elderly but did not include actual older adults as their research participants (we included the study in our analysis because the technologies under consideration were intended for use by older adults). The AI technologies or systems involved in these studies include social robots, e-learning, speech/image recognition, IoT and deep learning, AI platforms, intelligent wearable devices.

The studies were conducted in six different countries, including Canada, China, Qatar, Brazil, Romania, and Spain. Two studies were conducted in a home setting, one study was conducted in a preschool, and one study can be carried out in nursing homes. Many studies reported positive outcomes in favor of the AI technology system or apps being studied (or to-be-developed) or potential benefit of the targeted populations. However, since most studies did not involve real users, the levels of evidence of the studies in our final sample were low.

Three out of seven studies are designed for children, and another four studies are for the elderly, indicating the importance of addressing the needs of children and older adults in AI-related studies. One of these three studies also focused on people with disabilities.

3.4   Discussion

This paper conducted a systematic review of AI technologies used by marginalized populations and the possible factors that may affect their perception of AI technologies. Three databases (Web of Science, ACM digital library, IEEE explore) were selected for literature search and 151 non-duplicated articles came out. After title/abstract and full paper screening, 7 articles were included in the final review sample.

As aforementioned, AI has great potential for improving care of older adults, children, and people with disabilities. However, few systematic review efforts have been performed

Table 3.3 Coding framework

Author	Year	Study aim	Method	Participant	Country/Area	Setting	Related type of technology	Key finding
Ali et al. [1]	2021	To develop an app to help children learn new Arabic vocabularies interactively and independently using their smartphones or tablets	Observation, experiment, assessment survey	5 subjects	Qatar	Home	two AI platforms namely Clarifai and Houndify, image and speech recognition	The app enables the children to independently learn and find the objects around them along with their definitions
Cui et al. [8]	2021	To introduce an intelligent mobile body temperature monitoring and management system designed for nursing homes	n/a	n/a	China	Nursing home	Intelligent wearable devices	This system meets the basic requirements for nursing home use
Chuma et al. [6]	2020	To develop and evaluate an Internet of things (IoT) privacy–protected, fall-detection system for the elderly using radar sensors and deep learning	n/a	6 volunteers	Brazil	n/a	IoT, deep learning	The system obtained 99.9% accuracy in detecting falls by using the GoogleNet convolutional neural network

(continued)

Table 3.3 (continued)

Author	Year	Study aim	Method	Participant	Country/Area	Setting	Related type of technology	Key finding
Cooper et al. [7]	2020	To design a social assistive robot and companion	n/a	n/a	Spain	n/a	Social robot	The social robot was designed to address the needs of people with physical constraints and people in isolation due to infectious diseases. It contributes to improving hospital care and promoting independent living
Ionescu and Enescu [18]	2021	To review the use of the information technology tools in Romanian preschool education during the COVID-19 pandemic and investigate the difficulties encountered	Survey	More than 130	Romania	Preschool	E-learning	Preschool children, parents, and teachers can benefit from the inclusion of online tools

(continued)

Table 3.3 (continued)

Author	Year	Study aim	Method	Participant	Country/Area	Setting	Related type of technology	Key finding
Jaana and Paré [19]	2020	To compare mobile health technology use for Self-Tracking between older adults and the general adult population in Canada	Cross-sectional survey	4109 Canadian residents	Canada	n/a	Smart devices	Significant differences were found between the older adults and the general adult population in the use of smart technologies and the internet ($P < 0.001$)
Mahdi et al. [22]	2021	To design a social robot and design an online method to get stakeholders involved with user-centered design	Semi-structured interviews, online	9 families	Canada	Home	Social robot	Parents perceived the robot positively (especially from a therapeutic perspective)

on the adoption and perception of AI technologies by gender or race/ethnicity. This study aimed at addressing these gaps in literature. In particular, we focused on identifying existing work on AI used by certain marginalized populations and exploring factors that may have effects on how AI technologies are perceived.

Our findings indicate a huge gap for research on the use, adoption, and perception of AI technologies by communities that have previously experienced inequities in AI and COVID-19. We found that most AI-related research does not pay attention to these populations. This is evident in the following aspects. First, a large number of the articles in our initial searches had no empirical data; instead, they reported technical specifics (e.g., algorithms, framework, conceptual design, hardware design, and software design) or system architecture for designing AI systems. These findings show that research on AI by people with disabilities, older adults, and children are at the preliminary technical development stage, far from being evaluated at the user level. Second, among the studies that carried out experiments on real subjects, most of them were descriptive and exploratory by using a very limited size of samples. Third, most of the studies focused primarily on using AI to automatically collect data from users (e.g., sensors), and/or using the automatically collected data (e.g., videos of children with autism) to identify behavioral patterns of the targeted populations. Few focused on providing services to the targeted populations or providing information for their own use. This is important because previous literature highlights the significant concerns for marginalized populations using AI, who were often also more vulnerable during the pandemic [27].

This research gap could be attributed to the following reasons:

- As aforementioned, [30] pointed out that implementation of AI may not be easily transferable from country to country.
- Fordham et al. [13] addressed that in major disasters, marginalized populations tend to face disproportionate difficulties such as barriers to education, access to technologies, or other resources.
- Evaluation of the use of AI technologies and systems on some populations may need a restricted and time-consuming procedure for IRB approval, such as with children.

Key findings of these studies show positive results with the use of AI technologies or systems by marginalized populations. The seven studies included in our final sample were carried out in 6 different countries, which indicates the increased interest in this research area outside of the USA. It also calls for research of this kind in the USA. In addition, the researchers of these studies share different research backgrounds such as psychology, engineering, and computer science, indicating a great opportunity for interdisciplinary collaboration between researchers in multidisciplinary fields and disciplines.

This systematic review has its limitations. The selection of our initial search terms was not exhaustive. We used only "artificial intelligence" or "intelligent" as our AI-related search terms and may have missed systems or technologies that did not use these terms but used AI. The marginalized populations reviewed were not a comprehensive list and do not represent all marginalized populations globally. There is the potential for selection bias, whereby articles were unintentionally selected to support the belief of the researchers. To reduce the potential for this bias, four coders were used. Despite these limitations, our systematic review is valuable as it has identified existing work on AI with children, people with disabilities, and older adults.

3.5 Conclusion

The findings of this systematic review indicate the existing huge research gap in the field of AI, especially, how marginalized populations adopt AI technologies in their everyday lives, education, and work environment; and how they perceive AI technologies. With COVID-19, this research gap has been exacerbated. The findings also suggest more research is needed focusing on the adoption of AI technologies among marginalized populations or on the specific dimensions of vulnerability such as race/ethnicity, low-income populations, or gender.

Acknowledgements Bennett et al. (2020). NSF CONVERGE Working Group, COVID-19 Global Research Registry for Public Health and Social Sciences Technological Innovations in Response to COVID-19. This COVID-19 Working Group effort was supported by the National Science Foundation-funded Social Science Extreme Events Research (SSEER) Network and the CONVERGE facility at the Natural Hazards Center at the University of Colorado Boulder (NSF Award #1841338).

References

1. Ali, Z., Saleh, M., Al-Maadeed, S., Abou Elsaud, S., Khalifa, B., AlJa'am, J. M., & Massaro, D. (2021, April). Understand My World: An Interactive App for Children Learning Arabic Vocabulary. *In 2021 IEEE Global Engineering Education Conference (EDUCON)* (pp. 1143–1148). IEEE.
2. Atashzar, S. F., Carriere, J., & Tavakoli, M. (2021). How Can Intelligent Robots and Smart Mechatronic Modules Facilitate Remote Assessment, Assistance, and Rehabilitation for Isolated Adults with Neuro-Musculoskeletal Conditions? *Frontiers in Robotics and AI, 8*.
3. Bennett Gayle, D., X. Yuan, and T. Knight. (2021). Coronavirus Pandemic: The Use of Technology for Education, Employment, Livelihoods. *Journal Assistive Technology.* doi: https://doi.org/10.1080/10400435.2021.1980836. Epub ahead of print. PMID: 34813718.
4. Castets-Renard, C., & Fournier-Tombs, E. (2020). AI must be used responsibly with vulnerable populations. *Policy Options.* https://policyoptions.irpp.org/magazines/september-2020/ai-must-be-used-responsibly-with-vulnerable-populations/.
5. Chen, I. Y., Joshi, S., & Ghassemi, M. (2020). Treating health disparities with artificial intelligence. Nature medicine, 26(1), 16–17.

6. Chuma, E. L., Roger, L. L. B., De Oliveira, G. G., Iano, Y., & Pajuelo, D. (2020, December). Internet of things (IoT) privacy–protected, fall-detection system for the elderly using the radar sensors and deep learning. In *2020 IEEE International Smart Cities Conference (ISC2)* (pp. 1–4). IEEE.

7. Cooper, S., Di Fava, A., Vivas, C., Marchionni, L., & Ferro, F. (2020, August). ARI: The social assistive robot and companion. In *2020 29th IEEE International Conference on Robot and Human Interactive Communication (RO-MAN)* (pp. 745–751). IEEE.

8. Cui, C., Wang, C., Dong, X., & An, Z. (2021, April). Intelligent Mobile Body Temperature Monitoring & Management System For Nursing Home. In *2021 IEEE 6th International Conference on Computer and Communication Systems (ICCCS)* (pp. 812–818).

9. Dankwa-Mullan, I., Scheufele, E. L., Matheny, M. E., Quintana, Y., Chapman, W. W., Jackson, G., & South, B. R. (2021). A Proposed Framework on Integrating Health Equity and Racial Justice into the Artificial Intelligence Development Lifecycle. *Journal of Health Care for the Poor and Underserved, 32*(2), 300–317.

10. Ding, W., Nayak, J., Swapnarekha, H., Abraham, A., Naik, B., & Pelusi, D. (2021). Fusion of intelligent learning for COVID-19: A state-of-the-art review and analysis on real medical data. *Neurocomputing, 457*, 40–66.

11. Enache, B. A., Grigorescu, S. D., Adochiei, I. R., Eanche, C. D., Voiculescu, D. I., Argatu, V., ... & Stoica, C. (2020, June). Didactic Implementation of a Real-Time Biomonitoring Platform. In 2020 12th International Conference on Electronics, Computers and Artificial Intelligence (ECAI) (pp. 1–4). IEEE.

12. Evans, M. K. (2020). Covid's color line—infectious disease, inequity, and racial justice. New England Journal of Medicine, 383(5), 408–410.

13. Fordham, M., Lovekamp, W. E., Thomas, D. S., & Phillips, B. D. (2013). Understanding social vulnerability. Social vulnerability to disasters, 2, 1–29. CRC Press.

14. Gil-García, J. Ramón, Sharon S. Dawes, and Theresa A. Pardo. 2018. Digital government and public management research: finding the crossroads. Public Management Review 20 (5): 633–646.

15. Glauser, W. (2020). AI in health care: Improving outcomes or threatening equity? Can Med Assoc. DOI: https://doi.org/10.1503/cmaj.1095838.

16. Gozes, O., Frid-Adar, M., Sagie, N., Zhang, H., Ji, W., & Greenspan, H. (2020). Coronavirus detection and analysis on chest ct with deep learning. *arXiv preprint* arXiv:2004.02640.

17. Holstein, K., & Doroudi, S. (2021). Equity and Artificial Intelligence in Education: Will" AIEd" Amplify or Alleviate Inequities in Education? arXiv preprint arXiv:2104.12920.

18. Ionescu, V. M., & Enescu, F. M. (2021, July). Using information technology in Romanian preschool environment during the COVID-19 pandemic. In *2021 13th International Conference on Electronics, Computers and Artificial Intelligence (ECAI)* (pp. 1–6). IEEE.

19. Jaana, M., & Paré, G. (2020). Comparison of mobile health technology use for Self-Tracking between older adults and the general adult population in Canada: cross-sectional survey. *JMIR mHealth and uHealth, 8*(11), e24718.

20. Lee, N. T. (2018). Detecting racial bias in algorithms and machine learning. Journal of Information, Communication and Ethics in Society.

21. Luengo-Oroz, M., Bullock, J., Pham, K. H., Lam, C. S. N., & Luccioni, A. (2021). From Artificial Intelligence Bias to Inequality in the Time of COVID-19. IEEE Technology and Society Magazine, 40(1), 71–79.

22. Mahdi, H., Saleh, S., Sanoubari, E., & Dautenhahn, K. (2021, August). User-Centered Social Robot Design: Involving Children with Special Needs in an Online World. In 2021 30th *IEEE International Conference on Robot & Human Interactive Communication (RO-MAN)* (pp. 844–851). IEEE.

23. Mitra, R. (2020, May 7). *How Artificial Intelligence can help Vulnerable Populations during Pandemics.* Medium. https://becominghuman.ai/how-artificial-intelligence-can-help-vulnerable-populations-during-pandemics-aef530154cdd.
24. Moher, D., Altman, D. G., Liberati, A., & Tetzlaff, J. (2011). PRISMA statement. *Epidemiology, 22*(1), 128.
25. NITRD (2019). The National Artificial Intelligence Research and Development Strategic Plan: 2019 Update (nitrd.gov).
26. Office of the Privacy Commissioner of Canada. (2020, March 20). *Privacy and the COVID-19 outbreak.* https://www.priv.gc.ca/en/privacy-topics/health-genetic-and-other-body-information/health-emergencies/gd_covid_202003/.
27. Ondili, M. (2021, January 20). The Impact of the AI Divide on Vulnerable Groups. *Centre for Intellectual Property and Information Technology Law.* https://cipit.strathmore.edu/the-impact-of-the-ai-divide-on-vulnerable-groups/.
28. O'Neil, C. (2016). *Weapons of Math Destruction: How Big Data Increases Inequality and Threatens Democracy* (1st edition). Crown.
29. Pan, D., Sze, S., Minhas, J. S., Bangash, M. N., Pareek, N., Divall, P., ... & Pareek, M. (2020). The impact of ethnicity on clinical outcomes in COVID-19: a systematic review. *EClinicalMedicine, 23*, 100404.
30. Pedro, F., Subosa, M., Rivas, A., & Valverde, P. (2019). Artificial intelligence in education: Challenges and opportunities for sustainable development. Published in 2019 by the United Nations Educational, Scientific and Cultural Organization, 7, place de Fontenoy, 75352 Paris 07 SP, France.
31. Röösli, E., Rice, B., & Hernandez-Boussard, T. (2021). Bias at warp speed: how AI may contribute to the disparities gap in the time of COVID-19. *Journal of the American Medical Informatics Association, 28*(1), 190–192.
32. Shi, W., Peng, X., Liu, T., Cheng, Z., Lu, H., Yang, S., ... & Shan, F. (2021). A deep learning-based quantitative computed tomography model for predicting the severity of COVID-19: a retrospective study of 196 patients. *Annals of Translational Medicine, 9*(3).
33. Smith, M. J., Axler, R., Bean, S., Rudzicz, F., & Shaw, J. (2020). Four equity considerations for the use of artificial intelligence in public health. Bulletin of the World Health Organization, 98(4), 290.
34. Southgate, E. (2020). Artificial intelligence, ethics, equity and higher education: A 'beginning-of-the-discussion' paper. National Centre for Student Equity in Higher Education, Curtin University, and the University of Newcastle.
35. Suresh, H., & Guttag, J. V. (2019). A framework for understanding unintended consequences of machine learning. *arXiv preprint* arXiv:1901.10002.
36. Winter, J. S., & Davidson, E. (2019). Governance of artificial intelligence and personal health information. Digital policy, regulation and governance. VOL. 21 NO. 3 2019, pp. 280–29.
37. Wisner, B., & Blaikie, P. (2004). Cannon T& Davis I. *At risk: natural hazards, people's vulnerability and disasters.*
38. Young, M. M., Bullock, J. B., & Lecy, J. D. (2019). Artificial discretion as a tool of governance: a framework for understanding the impact of artificial intelligence on public administration. Perspectives on Public Management and Governance, 2(4), 301–313.

Xiaojun Yuan, Ph.D., is an Associate Professor in the College of Emergency Preparedness, Homeland Security, and Cybersecurity at the University at Albany, State University of New York. Her research interests include both Human Computer Interaction and Information Retrieval, with the focus on user interface design and evaluation and human information behavior.

She has received various grants and contracts, including from the Institute of Museum and Library Servcices, SUNY seed grant, Initiatives For Women Program at University at Albany, and New York State Education Department.

She published extensively in journals in information retrieval and human computer interaction (JASIS&T, IP&M, Journal of Documentation, etc.), and conferences in computer science and information science (ACM SIGIR, ACM SIGCHI, ACM CHIIR, ASIS&T, etc.).

Dr. Yuan received her Ph.D. from Rutgers University at the School of Communication and Information, and Ph.D. from Chinese Academy of Sciences in the Institute of Computing Technology. She received her M.S. in Statistics from Rutgers University and M.E. and B.E. in Computer Application from Xi'an University of Science & Technology in China. She serves as an Editorial Board Member of Aslib Journal of Information Management (AJIM), and a Board Member of the International Chinese Association of Human Computer Interaction. She is a member of the Association for Information Science and Technology (ASIS&T), the Association for Computing Machinery (ACM) and the Institute of Electrical and Electronics Engineers (IEEE).

DeeDee Bennett Gayle, Ph.D. is an Associate Professor in the College of Emergency Preparedness, Homeland Security, and Cybersecurity at the University at Albany, State University of New York. She broadly examines the influence and integration of advanced technologies on the practice of emergency management, and for use by vulnerable populations.

She has secured research grants and contracts, including from the National Science Foundation, Federal Emergency Management Agency, and the Department of Homeland Security. Her work is published in various journals, and she has presented at several conferences related to emergency management, disability, wireless technology, and future studies.

Dr. Bennett Gayle received her Ph.D. from Oklahoma State University in Fire and Emergency Management. She has a unique academic background having received both her M.S. in Public Policy and B.S. in Electrical Engineering from the Georgia Institute of Technology. She is an Advisory Board Member for the Institute for Diversity and Inclusion in Emergency Management (I-DIEM), a member of the Social Science Extreme Events Reconnaissance (SSEER) and Interdisciplinary Science Extreme Events Reconnaissance (ISEER), within the NSF-FUNDED CONVERGE initiative.

Thora Knight is an information science doctoral candidate in the College of Emergency Preparedness, Homeland Security, and Cybersecurity (CEHC) at the University at Albany, State University of New York. She is a JD/MBA graduate from the State University of New York at Buffalo and holds a BS in business concentrating on information systems from the University of Phoenix. Her research broadly focuses on understanding digital technologies' privacy and security implications. Her current dissertation study examines the role of regulation in mitigating privacy concerns in the context of online behavioral advertising.

Elisabeth Dubois is a Ph.D. Candidate in the College of Emergency Preparedness, Homeland Security, and Cybersecurity. She completed a B.S. in Digital Forensics and an MBA from the University at Albany, in 2018 and 2020 respectively. Her research examines risk management across numerous disciplines including communication, public administration, business, IT, cybersecurity, and education. Within these areas, she is interested in risk management among socially vulnerable populations amid COVID-19 and operational cyber risk management in a global environment.

Elisabeth's passion for this area of research stems from her volunteer work and serving as the Director of Marketing at The Global Child, a school and safe-house for former street-working children in Cambodia. In this role, she has engaged with and taught former street children, learning the realities of living in a post-war era and has been driven to investigate ways to address the inequities across sectors, both academically and practically. Using her educational background and passion for community endeavors, she has co-founded The Global CyberTech Exchange to help educate and train underserved populations for careers in cybersecurity. Her unique background is grounded in both her educational and research pursuits, as well as putting what she has learned into practice.

Mining the Health Information Needs of COVID-19 Patients Based on Social Q&A Community

4

Dan Wu and Le Ma

Abstract

COVID-19 has become a global pandemic, and COVID-19 patients are in a medical dilemma with no effective treatment and no effective drugs. The questions and answers in the social Q&A community can reveal the characteristics and evolution rules of the health information needs of COVID-19 patients. Using the Q&A data in Baidu Zhidao (https://zhidao.baidu.com/) as the research object, using the web crawlers to capture the data, automatic topic recognition on the acquired data by constructing an LDA topic model, exploring the content of COVID-19 patients' health information needs, and revealing the change rule of Q&A publication volume and health information need topics from the time dimension. Combining statistical information such as the number of answers, the number of likes, and the level of respondents, cluster analysis is used to reveal the changing rules of social characteristics and health information need topics. By analyzing the Q&A data on COVID-19 patients in Baidu Zhidao, it is found that the topic distribution of health information needs topic is relatively concentrated. Moreover, the number of Q&A and the types of health information needs to be changed in different development periods. There are differences in social characteristics that correspond to different topics of health information needs. Through in-depth analysis of the characteristics of health information needs of COVID-19 patients in the social Q&A community, on the one hand, it is beneficial for COVID-19 patients to obtain

D. Wu (✉) · L. Ma
School of Information Management, Wuhan University, Wuhan, China
e-mail: woodan@whu.edu.cn

Center for Studies of Human-Computer Interaction and User Behavior, Wuhan University, Wuhan, China

© The Author(s), under exclusive license to Springer Nature Switzerland AG 2023
X. Yuan et al. (eds.), *Social Vulnerability to COVID-19*, Synthesis Lectures
on Information Concepts, Retrieval, and Services,
https://doi.org/10.1007/978-3-031-06897-3_4

the required health information content timely. On the other hand, it is beneficial to optimize the community information display mechanism and improve the organization of information resources.

Keywords

COVID-19 • Social Q&A community • Health information need

4.1 Introduction

At present, COVID-19 epidemic prevention and control is becoming normal at home and abroad. According to the latest data on the website of the World Health Organization (WHO), more than 180 million cases of COVID-19 have been confirmed globally [32], and the need for health information about the symptoms, sequelae, and cure of COVID-19 patients has gradually become the focus of attention. In January 2021, a study in The Lancet, a leading medical journal, found that most patients discharged from hospital with COVID-19 still had a symptom six months after onset. At the same time, many research results show that novel coronavirus is likely to cause long-term effects on the human body, forming sequelae [28]. In February 2021, the World Health Organization's head of health care preparedness, Diaz, stated that patients recovering from an acute illness caused by COVID-19 would develop a different set of symptoms, known as "chronic novel coronavirus disease [29]." Therefore, it is urgent to explore the health information of COVID-19 patients further.

The rapid development of information technology drives the Internet to become an important channel for people to obtain health information. The Pew Research Center in the United States predicts that after the COVID-19 pandemic, by 2025, people will rely more on rapidly developing digital tools, such as telemedicine and virtual social networking [24]. In 2020, China's Internet industry played an active role in resisting the new crown pneumonia epidemic and the normalized prevention and control of the epidemic. Affected by the epidemic, Chinese netizens' demand for online medical services continues to increase, further promoting the digital transformation of China's medical industry. As of December 2020, the number of online medical users in China was 215 million, accounting for 21.7% of the total netizens [8]. Due to many users and simple information acquisition methods of social Q&A community (Q&A communities), it has gradually become one of the main ways for people to search for health information actively [19]. Compared with search engines and encyclopedia websites, social Q&A community can meet users' basic information search needs and provide users with social and emotional support [14]. However, under the new coronary pneumonia pandemic, there is insufficient research on the information needs of COVID-19 patients under the health topic. Most previous studies on the health needs of COVID-19 patients are empirical studies in the form of questionnaire surveys or interviews [1, 20], with a small amount of data involved in the studies.

This paper studies the question-and-answer data (Q&A data) on the health information needs of COVID-19 patients under the "Baidu Zhidao" community, firstly clarifies the topic of health information needs of COVID-19 patients, and then further explores the law of changes in the topic of health information needs of COVID-19 patients over time, and finally analyzes users' social attributes and theme COVID-19 patients health information demand. Based on clarifying the content of the health information needs of COVID-19 patients, the above analysis can further provide a positive reference for the socialized Q&A community to improve the service level.

4.2   Related Work

4.2.1   Research on Health Information Needs

Health information refers to people's physical and mental health, disease, nutrition, and health maintenance, which can guide personal health or clinical behavior. Users will choose to search for various health information through TV, Internet, books, and other ways due to their special health conditions, need to further consult a doctor about personal health problems or know about nutrition, exercise, and weight control [13, 21]. Nowadays, with the rapid development of the Internet, more and more users choose to search for health information through the network. In this process, the demand for online search for health information has increasingly become a research hotspot. Cleveland et al. [9] investigated the use of Chinese online health information resources. They found that the most concerning health information were nutrition information and information about specific diseases and that users were highly motivated to learn and search for health information online. Pieper et al. [26] investigated the health information needs of German patients, relatives, and the general population who can speak German or English during 2000–2012 through systematic review method, and found that the younger the age. The shorter the course of illness, the worse the health status, the greater the tendency of anxiety and depression, and the correspondingly high demand for health information. Vetsch et al. [30] through center methods describe the information needs of breast outpatients, through questionnaire and interview to understand cancer survivors and their families of information needs, and unmet needs and the connection between the clinical and sociodemographic characteristics, studies have shown that survivors do not meet the needs of medical information, and relatives in the information needs of life. As a relative, overall health status was reduced, the perceived risk of late effects was high, and a significant correlation was found between the degree of anxiety and depression and unmet health information needs. Cappelletti et al. [6] used a questionnaire to explore the impact of two chronic diseases, hypertension, and coronary disease, on patients' health information needs. The survey found that over time, medical experts were considered trustworthy sources of health information. A semi-structured telephone interview and patient focus group discussion were conducted to understand the information needs of patients starting oral antidiabetic therapy while studying the opportunities for pharmacies to provide information to patients with diabetes, and found that general practitioners were unable to meet the health information needs of patients. Jean et al. [16] investigated the information needs, information needs, and use of diabetic patients. Questionnaires, interviews, and card classification were used to allow participants to judge the usefulness of diabetes-related information from different sources and types at different time points.

4.2.2 Research on Health Information Needs and User Behavior in Socialized Q&A Community

The social Q&A community is an open service community for all network users. Participants ask questions purposefully. The community provides certain rewards based on the answer quality of the answerers, thereby motivating users to participate in it actively [27]. The Q&A knowledge base established by the health Q&A community is incorporated into the information search source, improving the accuracy, diversity, and accessibility of knowledge acquisition, better serving users' health information search behavior, and meeting users' health information needs.

Research on users' health information needs to be based on data from socialized Q&A community mainly focuses on the following topics: (1) Types of health information needs. Oh et al. [22] used cancer-related questioning data as the research object and found that the information needs for cancer include six categories: demography, cognition, emotion, society, context, and technology. The research of Zhan et al. [35] took Zika virus-related Q&A data as the research object and divided the information needs about Zika virus into causes, symptoms, transmission, risks, protection, monitoring, and prevention. Bowler et al. [5] obtained user data from the Yahoo Q&A community and divided the health information needs into five categories: seeking information, seeking emotional support, seeking communication, seeking self-expression, seeking help to complete tasks, and dividing them into diagnosis, treatment, or intervention, etc. 11 sub-categories. (2) Differences in health information needs of different users. With diabetic patients as interviewers, Armstrong and Powell [1] learned about the hot topics that patients actively participated in discussions in the online health community. Zhang and Zhao [34] analyzed the diabetes-related Q&A records in the Yahoo Q&A community and revealed twelve categories of health topics that are of concern to people with diabetes. Zhao et al. [36] collected Q&A data related to depression on Zhihu, a community Q&A community. They found that the symptoms of depression and social activities were the health information content that users paid the most attention to, while health information such as prevention and medical insurance had a low demand for attention. Bahng and Lee [2] analyzed the Q&A data of deaf and hard-of-hearing patients on social Q&A community. They found that the patients were concerned about sudden hearing loss, tinnitus, noise-induced hearing loss, hearing AIDS, dizziness, and diseases and symptoms. Pfeil and Zaphiris [25] used social network analysis and found that empathic elderly communities where users with similar life backgrounds provide information exchange and emotional support have closer social connections than non-empathetic elderly online communities. Kanayama [18] analyzed the information texts released by the online community for the elderly and found that the elderly in the community expressed their concerns about the weather, health, and environmental information. They were more rigorous and polite in the process of exchanging information.

The current research is to study the behavior characteristics of users of social Q&A community from various angles. Gazan [12] proposed and explained the concept of micro-collaboration and pointed out that the main interaction method of social Q&A community users is micro-collaboration. Furtado et al. [10] performed cluster analysis on user behavior data (such as the number of questions answered, the number of questions asked, the number of likes, etc.) of users on the social Q&A community to find out the experts and active contributors with high motivation in different subject areas, which is conducive to assigning new questions to potential respondents. In the social Q&A community, user participation behaviors are mainly divided into active participation type, that is, they will actively participate in the Q&A interaction and contribute their own experience and knowledge,the other is the passive response type, which is users who only browse without responding and feedback. The study found that the more contributions, The larger the users, the more rewards they get [7, 11]. Moreover, starting from social exchange theory and social center theory, Jin et al. [17] studied the motivation of users to share knowledge on social Q&A community and found that users' self-expression, self-cognition, and social learning play an important role in it.

4.3 Research Design

4.3.1 Data Source and Collection

Baidu Zhidao (https://zhidao.baidu.com/) is based on search and realizes interactive knowledge Q&A sharing community. By 2020, the number of users of Baidu Knows reached 900 million, and the cumulative number of Q&A reached 3.83 billion. The community has significant advantages in terms of traffic, the number of users, resources, services, and technologies, and ranks first in the Chinese market regarding user awareness and usage penetration [15]. Baidu Zhidao is well aware of the high usage rate among Chinese netizens, which makes it one of China's social Q&A community representatives and can reflect the information demand of the Internet.

Currently, the number of COVID-19 diagnoses continues to increase, but there is little research on the health information needs of COVID-19 patients. Because of this, the Q&A data known to Baidu Zhidao are selected as the research object in this paper, and the

question	questionTime	authorName	authorLevel	answerTime	acceptCount	likeCount	content
除了经典症状外，COVID-19	2021-01-19	永远在旅游的i	3	2021-01-21			展开全部新型冠状病毒在中国武汉首次出现时，不知道有什么异常表现。
除了经典症状外，COVID-19	2021-01-19	烧烤味吻你我	1	2021-04-01			展开全部一些COVID-19患者表现出较少的典型症状，包括恶心、腹泻、:
除了经典症状外，COVID-19	2021-01-19	心理疏导学生	0	2021-03-31	26	191	展开全部首先这个时候可能会出现发热的症状，并且很多时候你会出现:
除了经典症状外，COVID-19	2021-01-19	贪睡的凤凤	7	2021-03-31			展开全部COVID-19除了一些比较典型的发烧头痛咳嗽之类的症状之外，:
COVID-19如何影响心理健康	2021-01-19	永远在旅游的i	3	2021-03-29			展开全部冠状病毒的蔓延让人们心里充满恐慌，宅在家的时候，看到新i
COVID-19如何影响心理健康	2021-01-19	百度网友0c77c	5	2021-03-29			展开全部这种疾病对所有的影响都非常大，他们不仅是身体上还有心理.
COVID-19如何影响心理健康	2021-01-19	你真的好啊0001	1	2021-03-29			展开全部新冠在初期，因为人们缺乏对它的认识，其传播速度过快，所i
COVID-19如何影响心理健康	2021-01-19	默默无闻奇观i	0	2021-03-29	0	6	展开全部新冠在初期，因为人们缺乏对它的认识，其传播速度过快，所i
COVID-19如何影响心理健康	2021-01-19	乐乐在此呢	0	2021-03-22	0	75	展开全部恐惧和压力，我们每个人面临问题都有不同压力，而且还有后i

Fig. 4.1 Data collection results

identified search terms are expressed in Chinese as 新冠患者(COVID-19 patients)、新冠病人(COVID-19 patients)、新冠肺炎患者(COVID-19 patients)、新冠病毒(COVID-19)、新型肺炎(COVID-19)、新型冠状病毒(COVID-19)、新冠疫情(COVID-19 epidemic)、新冠疫区(COVID-19 epidemic area)、新冠隔离(COVID-19 quarantine). On March 12, 2020, the Director-General of the World Health Organization, Tedros, declared the COVID-19 outbreak a global pandemic. Therefore, in this study, the retrieval time is limited to March 12, 2020, solstice, March 12, 2021, to obtain the health information needs of COVID-19 patients in the year when COVID-19 entered the global pandemic state. In the search process, use the advanced search function provided by Baidu Zhidao itself to limit the search terms and search time in the search results.

In this paper, the web crawler written in Python language is used to obtain all Q&A data, and the collected data were screened to remove the data, not in line with the research topic. Finally, 18,842 Q&A data known to Baidu were obtained (including 4148 Q&A data and 14,694 answer data). At the same time, the relevant data of the respondents (including grade, adoption number, thumb up number, etc.) are also obtained. These data are stored in Excel files, and the final data collection results are shown in the Fig. 4.1.

4.3.2 Data Processing

In this study, the stopwords database of Harbin Institute of Technology (https://github.com/goto456/stopwords/blob/master/hit_stopwords.txt) was used, and a total of 676 stopwords were used. Stopping words are words with no real meaning, such as primary and secondary words, prepositions, conjunctions, modal words, etc. The obtained data is imported into the stop list through code. After that, word segmentation is carried out on the Q&A data. Chinese word segmentation includes the word segmentation method of dictionary and the word segmentation method based on statistical segmentation. This study is based on the Jiebar package developed by R language to achieve word segmentation. In word segmentation, stop words are removed to ensure the accurate extraction of text features. Then, the data type afterword segmentation is converted into the corpus form, and the corpus is constructed into a document entry matrix (DTM) by the method DocumentTermMatrix (). DTM is a two-dimensional matrix in which the first row represents all the feature words in the corpus, the first column represents the serial number of the documents asked by users, and the value of the matrix represents the frequency of feature words in each document. Due to a large number of DTM dimensions initially generated, in order to improve the running speed and clustering accuracy of subsequent algorithms, feature screening and extraction of the initial DTM should be carried out through principal component analysis (PCA), singular value decomposition (SVD) and manual feature screening.

4.3.3 LDA Topic Model

LDA topic is proposed based on the semantic analysis model, which is a probabilistic algorithm. It is a three-layer Bayesian model, including document, topic, and lexical items. It is essentially a clustering algorithm [33]. The important parameter of the model modeling is the number of topics, which needs to be specified before modeling to guide the algorithm to assign topics to documents. The number of topics has a great influence on the results of model training. Therefore, in this study, the perplexity index is used to determine the optimal number of topics [4]. The perplexity is the topic model obtained by training. For document D, the degree of uncertainty of which topic should belong to is numerically compared with all words. The reciprocals of the geometric mean of the item probabilities are equal. When other conditions remain unchanged, the perplexity will decrease as the number of topics increases. Therefore, the number of topics corresponding to the inflection point in the perplexity-topic number curve is generally taken as the optimal number of topics.

4.3.4 Theme Coding

The content analysis method is one of the common methods used to study the social Q&A community, which can systematically summarize the basic characteristics of many questions. The information needs of users in the social Q&A community influence their questioning and answering behaviors. Due to the particularity of users' health information needs in the social Q&A community, some studies have divided health information needs into disease-related and non-disease-related categories. Disease-related information needs include etiology, diagnosis, treatment, etc., while non-disease-related information needs include spiritual and social aspects [27]. In order to further simplify the classification, three categories of "infectivity", "emotional support," and "social support" were added to measure the impact of the large-scale outbreak of COVID-19 and the impact on users' psychology during the special period of the epidemic. In this paper, the query data was manually coded by a three-level coding method [23], and second-level coding subject words represented the health information needs of users. Researchers checked each piece of data. Cohen Kappa value was used to quantify the consistency of the data set coding results, and the results showed that Cohen Kappa value was 0.731 (P > 0.001), indicating that the classification results are reliable [31]. The specific contents of the coding table are shown in Table 4.1.

Table 4.1 Coding table of topics involved in the data of questions from the social Q&A community about COVID-19

Third-level coding	Second-level coding	First-level coding	Description
Disease directly related	Causes, symptoms and manifestations	Cause	Ask about the cause of COVID-19
		Symptoms and manifestations	Ask about basic symptoms and basic knowledge related to COVID-19
	Inspection and diagnosis	Detect	Ask if you need a nucleic acid test
		Confirmed diagnosis	Ask if you have been infected with the COVID-19
	Treatment	Treatment effect	Ask about existing effective treatments
		Other treatment methods	Ask if there are any treatment drugs
	Prevention	Preventive measures	Ask how to prevent and avoid new coronary pneumonia
		Prevention knowledge	Ask about knowledge related to the prevention of COVID-19
	Infectious	Infected people	Ask about the number of confirmed cases of COVID-19, the number of deaths, asymptomatic infections, etc
		Region	Ask about the spread of the COVID-19 epidemic in the world

(continued)

Table 4.1 (continued)

Third-level coding	Second-level coding	First-level coding	Description
		Virus	Ask about the variants of COVID-19
Not directly related to disease	Emotional support	Spiritual support	Seeking encouragement and comfort from others, etc
	Social support	Social life	Ask about the impact of patients with new coronary pneumonia in social life

4.4 Analysis of the Results of COVID-19 Patients' Health Information Needs in Social Q&A Community

4.4.1 Basic Situation of Q&A Data

In this paper, through manual screening, Excel screening, and reduplication function, the original question data collected were deleted irrelevant data and duplicate data, and 18,842 pieces of Q&A data were obtained. On this basis, the COVID-19 special word list, the custom stop word list, the synonym list, and the regional replacement word list were established, and the data preprocessing was completed by using Chinese word segmentation, stop word and word frequency statistics techniques. In order to study the focus of COVID-19 patients' health information needs, this paper presents high-frequency subject words through a word cloud map, which contains the top 100 subject words. Figure 4.2 shows that "covid-19", "coronavirus", "patient", "epidemic" and "case" are the focus of the re-mutation points of patients with coronavirus, while "vaccine", "prevention", "isolation" and "mutations" are the key health information issues of patients with coronavirus. At the same time, the keywords "China", "United States", "Japan," and other countries can be seen, indicating that COVID-19 patients in different countries and regions are concerned about the cure of COVID-19.

In this study, the number of confirmed COVID-19 cases globally published by WHO and the number of questions from COVID-19 patients at different times was combined to draw the distribution of health information demand data of COVID-19 patients during the period, as shown in Fig. 4.3. As can be seen from the figure, according to the number of COVID-19 confirmed cases published by WHO, the development of COVID-19 can be divided into three main stages according to the time nodes of the development of the epidemic: The first phase is the initial global outbreak of COVID-19 (2020.03–2020.05), the second phase is the rapid global rise of COVID-19 (2020.06–2020.11), and the third phase is the global vaccination phase (2020.12–2021.03). As shown in Fig. 4.3,

Fig. 4.2 Topic word cloud diagram

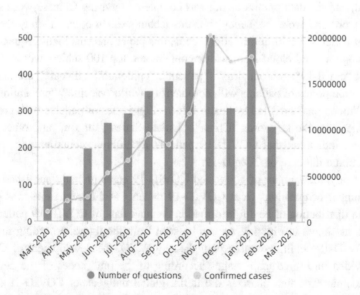

Fig. 4.3 Distribution of health information needs data period

in the early stage of the global COVID-19 outbreak, COVID-19 patients had less demand for health information, but COVID-19 entered the rising stage, the demand for health information COVID-19 patients increased significantly. The level of demand for health information from COVID-19 patients has risen again during the mass vaccination campaigns that began in many countries worldwide in December 2020. It is higher than at the beginning of the COVID-19 outbreak, indicating increased expectations for effective treatment for COVID-19 patients after the advent of the COVID-19 vaccine. Proactively search for reliable health information. It can also be seen from the figure that the level of health information demand of COVID-19 patients is positively correlated with the development stage of the global COVID-19 epidemic; that is, with the increasing number of confirmed cases of COVID-19 globally, the level of health information demand will increase.

4.4.2   The Topic of Health Information Needs Based on COVID-19 Questioning Data

The topics for the health information need of COVID-19 patients in the social Q&A community are shown in Fig. 4.4. The topics are divided into seven categories: topic 1 is Causes, symptoms, and manifestations; topic 2 is inspection and diagnosis; topic 3 is treatment; topic 4 is prevention; topic 5 is infectious; topic 6 is emotional support; topic 7 is social support. Furthermore, the highest proportion of users who asked about

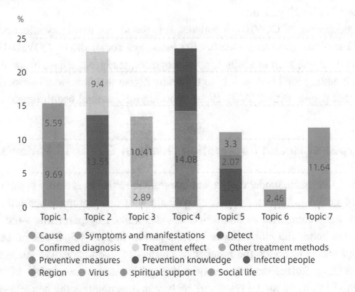

Fig. 4.4 The topic of health information needs

COVID-19 prevention is 23.38%, with 14.08% of users referring to prevention methods and advising on effective methods or measures to prevent COVID-19. Secondly, questions about examination and diagnosis information accounted for 22.95%, and queries about testing accounted for 13.55%, significantly higher than those about the diagnosis. It can be seen that users are most concerned about testing whether they are infected with COVID-19, and they have a strong willingness to test COVID-19.

The proportion of causes, symptoms, and manifestations is 15.28%, among which the proportion of cases is 9.69%. Questions about treatment accounted for 13.31% and generally occurred simultaneously as the cause, symptoms, and presentation, with users asking about COVID-19 symptoms followed by further questions about effective treatments for COVID-19. It can be seen that users are more concerned about the causes of COVID-19 infection, such as "which groups are more likely to be infected with COVID-19 and what are the current factors of COVID-19 infection"; 5.59% of users' questions are related to symptoms and manifestations, indicating that some patients suspected of COVID-19 are able to understand the symptoms of COVID-19 proactively.

The infectivity of COVID-19 is unique, so it has attracted the attention of users, especially as the COVID-19 epidemic is still raging around the world, with new confirmed cases and deaths every day, and the spread of COVID-19 is becoming more and more widespread, so the proportion of questions is 5.59%. In addition, the epidemic situation is different in each region, and the means of epidemic control are also different. The number of questions about the region is also 2.07%. The proportion of the questions of COVID-19 is higher than that of the region, accounting for 3.3%, indicating that the variation, transmission speed, and transmission mode of novel coronavirus are also one of the topics of concern for COVID-19 patients.

In the questioning of COVID-19 patients, it is found that users pay attention to social support information, especially whether the individual social life of COVID-19 infected patients has changed, such as social life, employment, study, etc. These become the focus of users, accounting for 11.64%. It is significantly higher than emotional support (2.46%), indicating that people with COVID-19 are ignoring their mental health issues.

4.4.3   Types of Health Information Needs of COVID-19 Patients

Analyze the LDA theme model results and integrate the existing category system to clarify the types of health information needs of patients with new coronary pneumonia in this article. The main categories in this article are determined by perplexity—the lower the perplexity, the better the clustering effect. As shown in Fig. 4.5, it can be seen that it is more appropriate to divide the number of clusters of five topics. The same method is used to determine further the number of subcategories under the number of categories. Combine the LDA topic model results to analyze and summarize the demand categories, and finally determine that There are five major categories and 12 subcategories.

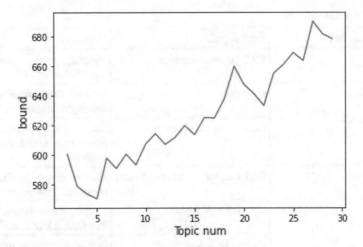

Fig. 4.5 Schema of degree of confusion

It can be seen from Table 4.2 that the health information needs of COVID-19 patients in the social Q&A community have the following characteristics: (1) In terms of the number of demands, users have various categories of demands and used many keywords, especially "clinical manifestation of disease", "examination and diagnosis" and "prevention" related to disease knowledge. It can be seen that COVID-19 patients attach great importance to the function of medical science popularization and health Q&A in social Q&A community. (2) In terms of the type of needs, COVID-19 patients in the social Q&A

Table 4.2 Classification of health information needs of COVID-19 patients

Major categories	Subcategories	Keywords
C1 etiology, symptoms and manifestations	C1.1 basic features of disease	Coronavirus, emergence, COVID-19, infectious disease
	C1.2 mode of disease transmission	Air, freezing, evidence
	C1.3 clinical manifestations of the disease	Death, respiratory, pulmonary, failure, incubation period, control
C2 test and diagnosis	C2.1 inspection, examination	Nucleic acid, positive, test
	C2.2 symptoms	Fever, cough, body temperature
C3 prevention	C3.1 prevention knowledge	Mask, disinfection, queuing, protection

(continued)

Table 4.2 (continued)

Major categories	Subcategories	Keywords
	C3.2 preventive measures	Vaccine, spacing, isolation, inoculation
C4 treatment	C4.1 drug	Oral solution, antibiotics, ShuangHuangLian
	C4.2 treatment approach	Isolation, ICU, medical observation, ventilator, plasma, antibody
C5 infectious	C5.1 number of confirmed cases	Number of new, death, cumulative, severe, mild disease
	C5.2 infected region	Country, transmission, Japan, The United States, India, China
	C5.3 virus evolution	Mutation, research, discovery

community mainly focus on the symptoms, modes of transmission, and clinical manifestations of the disease. (3) From the perspective of the demand cycle, COVID-19 patients in the social Q&A community have a strong continuity in their demand for health information. From "infectivity", it can be seen that they pay attention to the affected regions and populations in China and pay attention to the affected regions and populations in other countries in the world. Also, focus on the particular case of the novel coronavirus that has changed over time.

4.4.4   Changes in Health Information Needs of COVID-19 Patients in Different Periods

As shown in Fig. 4.6, there are significant differences in health information needs that COVID-19 patients focused on at different periods. (1) The difference in the number of topics paid attention to in different periods. In the early days of the COVID-19 global outbreak, the first low point in attention for the five topics was recorded, with the lowest number of attentions recorded in May 2020. During the rising stage of the COVID-19 outbreak, the amount of attention on the five topics increased significantly, and the first peak appeared, indicating that with the rapid development of the COVID-19 epidemic, users' attention on COVID-19 increased significantly. The social Q&A community is used to consult about COVID-19, but users' attention on different topics is significantly different. The topic of etiology, symptoms, and manifestations received the highest amount of attention, followed by infectivity, and the topic of treatment received the lowest amount of attention. In the new stage in the champions league vaccination, five topics appeared at the second peak. However, the five topics of concern both in the overall quantity are

count

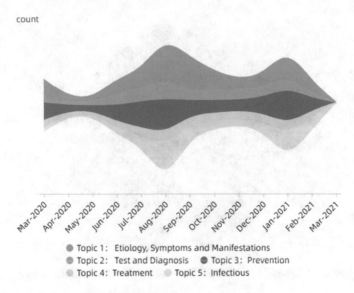

● Topic 1: Etiology, Symptoms and Manifestations
● Topic 2: Test and Diagnosis   ● Topic 3: Prevention
● Topic 4: Treatment   ● Topic 5: Infectious

Fig. 4.6 Changes in themes in different periods

lower than the first peak. It has changed with time, and COVID-19 patients actively seek COVID-19 related health information needs enthusiasm reduced, and explain new crown's mass vaccination activities. It has reduced people's fear of COVID-19 and made them more confident in preventing and controlling the epidemic. (2) The changes of different themes in different periods. The number of concerns of topic 1 is significantly higher than that of other topics in different periods, indicating that this topic is the main concern of COVID-19 patients in different stages of the development of COVID-19. The topic with the least amount of attention is Topic 4, which shows that in different periods, patients with COVID-19 are less proactive in seeking treatment when no effective treatment is available. In different stages of the development of COVID-19, the number of attentions paid to prevention is also second only to the topic of etiology, symptoms, and manifestations. It can be seen that understanding prevention knowledge or prevention means is one of the important contents in the COVID-19 epidemic period, and people take the initiative to consult the contents related to prevention to ensure effective prevention work.

4.4.5 Differences in Health Information Needs of Covid-19 Patients Among Users of Different Levels in the Social Q&A Community

In this paper, according to Baidu Zhidao, the grading system, namely, knows the grade [3]. The rating is based on the user's actions, such as answering, interacting, and completing tasks. According to adoption rates in the rating index, the user data collected in the

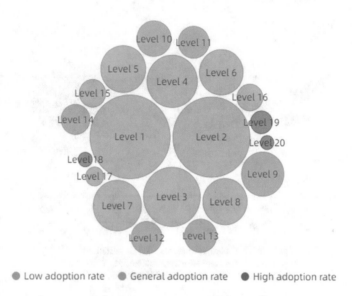

● Low adoption rate   ● General adoption rate   ● High adoption rate

Fig. 4.7 Level distribution of answerers

study are divided into three categories: ≤50% as low adoption rates, 52–78% as average adoption rates, and ≥82% as high adoption rates. As shown in Fig. 4.7, among the 20 grades, the number of low adoption rates is the largest, the number of ordinary adoption rates is small, and the number of high adoption rates is the least. It indicates that in the questions and answers about COVID-19 patients, most respondents' answers are not adopted by other users.

Based on the above three types of respondents, this paper analyzes the changes in the theme of health information needs of COVID-19 patients at different stages of the development of the COVID-19 epidemic. As shown in Fig. 4.8, (1) From the perspective of topic type, in the first stage, i.e., the initial stage of the global outbreak of COVID-19, topic one, "Etiology, Symptoms and manifestations" and topic three, "prevention," are the main topics for different levels of respondents. It indicates that at the early stage of the global outbreak of COVID-19, users want to learn more about the pathogenesis, pathological characteristics, and effective prevention methods of COVID-19 to improve their ability to respond to COVID-19. In the second phase, which is the rising COVID-19 epidemic, the number of users who participated in the second topic, "Detection and diagnosis", based on the focus on topic one and topic three, increased significantly, reflecting that as the COVID-19 epidemic continues to spread around the world, users want to learn how to detect COVID-19 infection and confirm COVID-19 proactively. In the third stage, the new stage of vaccination, theme five, "contagious," also become one of the main questions of content, visible with the development of the new champions league vaccination campaign, the user for the spread of COVID-19 districts, confirmed cases and deaths have

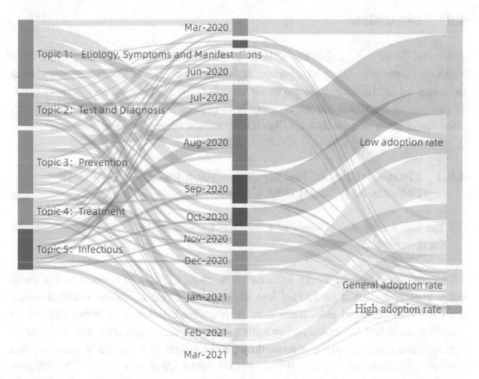

Fig. 4.8 Changes in the topic of information needs of users of different levels

also increased, especially will be coronavirus variant appeared at this stage, COVID-19 patients are paying more attention to the transmission of novel coronavirus. (2) From the perspective of different levels of respondents, the participation rate of respondents with a low adoption rate is very high in each topic. Especially in the rising phase of the COVID-19 outbreak, the participation rate of respondents with a low adoption rate is significantly increased, and the number of respondents is much higher than in other states. The general adoption rate respondents had the largest number of participants in the rising phase of COVID-19 and mainly participated in the answers on topic two and topic three. Respondents with a high adoption rate actively participated in all stages of questioning, and there is no significant difference in the number of participants for each topic. (3) From the point of different periods, COVID-19 patients' health information demand is mainly concentrated in the second phase, the outbreak of COVID-19 rising stage and the third phase, the new stage in the champions league vaccination, and these two stages, different levels of respondent participation are many, but the differences between different levels of respondents to participate in the theme of the content.

4.5   Discussion

The social Q&A website provides COVID-19 patients with an online health information exchange community. Users who ask and answer in the process can put forward and meet the needs of information. This paper selects the Q&A data on patients with COVID-19 patients on the Baidu Zhidao community to explore the type and evolution of health information needs of patients with COVID-19 patients during the global pandemic.

4.5.1   Characteristics of Health Information Needs of COVID-19 Patients in a Social Q&A Community

(1) The health information needs of COVID-19 patients in the social Q&A community are diverse.

This study found that the health information needs of COVID-19 patients in the social Q&A community could be divided into disease-related and non-disease-related needs. On the disease demand side, users want to learn basic information about COVID-19 infection, including the cause and symptoms of the disease, methods, and methods of detection and diagnosis of the disease, drugs, and methods of treatment, knowledge, and methods of prevention, and infertility, from the social Q&A community. The different categories are further subdivided into one to three sub-categories, illustrating the diversity of users' health information needs. In terms of non-disease needs, users want spiritual support through the social Q&A community, and some of the questions are directly related to psychological problems. During the COVID-19 epidemic, users need to seek reliable spiritual comfort from the social Q&A community to overcome this special period. Other users want further to discuss the impact of COVID-19 on their social life, whether they will be discriminated against in society after being infected with COVID-19, and whether their daily social interactions will change significantly, etc. The above different categories reflect the diversity of health information needs.

(2) The types of health information need of COVID-19 patients in social Q&A community are different at different times.

COVID-19 is a sudden infectious disease, which spreads differently at different times. Users' health information needs reflect the different stages of the development of the epidemic. Based on the volume of user posts and global COVID-19 confirmation, the outbreak can be divided into the initial phase of the COVID-19 outbreak, the rising phase of the COVID-19 outbreak, and the period of COVID-19 vaccination. In the early stages of the COVID-19 outbreak, the number of posts and types of health information needs of users is low; In the rising phase of COVID-19, the number of posts by users has increased significantly, and the demand for health information has increased significantly. During the period of COVID-19 vaccination, users' demand for health information gradually declined.

(3) The unique health information needs of COVID-19 patients in the social Q&A community are infectious.

The research of this paper found that under the COVID-19 epidemic situation, the health information demand related to infectious diseases in the question-and-answer data on the social Q&A community is one of the main contents that users' attention. As can be seen from the keywords of infectious disease, users have asked about the number of confirmed cases, the affected areas, and the variation of the virus many times on the social question-and-answer community. As the COVID-19 epidemic is still raging around the world, the number of deaths and confirmed cases are constantly at a new high, which has attracted wide attention.

4.5.2 Differences in Health Information Needs of COVID-19 Patients with Different Social Attributes in Social Q&A Community

Through the analysis, it is found that users of different social levels also have significant differences in their needs for health information over time. In the first phase of the rapid spread after the COVID-19 pandemic was declared, people with low adoption rates actively participated in question-and-answer interactions about the causes, symptoms, and manifestations of COVID-19. With the recurrence of the epidemic, there will be another peak of question-and-answer interaction in January 2020, during which the demand for health information will focus on prevention knowledge and prevention methods.

4.6 Conclusion

This paper explores the health information needs of COVID-19 patients during the COVID-19 pandemic based on social question-and-answer communities through the research steps of data collection, data processing, LDA thematic model, theme coding and need identification. After statistical analysis of the Q&A data, the types of health information needs are determined. Combined with different periods, the changes of health information need in different periods are analyzed to reveal the differences in health information needs of users with different social attributes in different periods. In theory, it provides a basis for further understanding the types and changes of health information needs of COVID-19 patients in the social Q&A community under the continuous spread of the COVID-19 epidemic. In practical application, the research results of this paper can provide certain support for improving the health information service level of health websites, such as information resource organization, ranking of health information search results, optimization of health information display mechanism, etc.

There are also some deficiencies and limitations in the study of this paper. As the selected Q&A data are the data of a certain period on a single community, it is difficult to conduct comparative analysis on the health information needs of COVID-19 patients in different periods of COVID-19 on different social Q&A community. Other algorithms can do future research in different social Q&A community to obtain the Q&A data of patients with COVID-19 patients, and clustering analysis and themes of these data are extracted, which can be more comprehensive summarize and analyze during the pandemic—the evolution of the patient's health information needs and interactive evolution regularity of health information needs of users.

Acknowledgements The authors would like to acknowledge the help and support of project of the National Social Science Foundation of China, "Research on Multilingual Information Organization and Retrieval for the Integration of Resources of Three Public Digital Cultural Projects" (Item NO.: 19ZDA341).

References

1. Armstrong, N., and Powell, J. (2009). Patient perspectives on health advice posted on Internet discussion boards: a qualitative study. *Health Expectations*, 12(3), 313–320. DOI: https://doi.org/10.1111/j.1369-7625.2009.00543.x.
2. Bahng, J., and Lee, C. H. (2020). Topic Modeling for Analyzing Patients' Perceptions and Concerns of Hearing Loss on Social Q&A Sites: Incorporating Patients' Perspective. *International Journal of Environmental Research and Public Health*, 17(17), 6209. DOI: https://doi.org/10.3390/ijerph17176209.
3. Baidu Zhidao (2021), "Baidu Zhidao how to upgrade", Retrieved from http://help.baidu.com/question?prod_id=9&class=338&id=1507.
4. Blei, D. M., Ng, A. Y., and Jordan, M. I. (2003). Latent dirichlet allocation. the Journal of Machine Learning Research, 3, 993–1022. Retrieved from https://www.jmlr.org/papers/volume3/blei03a/blei03a.pdf?TB_iframe=true&width=370.8&height=658.8.
5. Bowler, L., Oh, J. S., He, D., Mattern, E., and Jeng, W. (2012). Eating disorder questions in Yahoo! Answers: Information, conversation, or reflection?. *Proceedings of the American Society for Information Science and Technology*, 49(1), 1–11. DOI: https://doi.org/10.1002/meet.14504901052.
6. Cappelletti, E., D'addario, M., Steca, P., Sarini, M., Greco, A., Monzani, D., and Pancani, L. (2013). Evolution of information needs in patients affected by coronaropathy and hypertension. In *Proc. the 27th Conference of the European Health Psychology Society*. Taylor and Francis, London, the United Kingdom, 182–182. DOI: https://doi.org/10.1080/08870446.2013.810851.

7. Changping, H., and Li, W. (2014). The influencing factors of knowledge sharing behavior on college students in virtual communities. *In Proc. In 2014 13th International Symposium on Distributed Computing and Applications to Business*, Engineering and Science. IEEE, Xi'an, China, 239–242. DOI: https://doi.org/10.1109/DCABES.2014.52.

8. China Internet Network Information Center (CNNIC) (2021), The 47th Statistical Report on Internet Development in China, Retrieved from http://www.gov.cn/xinwen/2021-02/03/content_5584518.htm.

9. Cleveland, A. D., Della Pan, C. J., Xinyu, Y., Philbrick, J., & O'Neill, M. (2008). Analysis of the health information needs and health related Internet usage of a Chinese population in the United States. Library and Information Service, 52(03), 112–116. Retrieved from https://covid19.who.int/info http://www.lis.ac.cn/CN/.

10. Furtado, A., Oliveira, N., and Andrade, N. (2014). A case study of contributor behavior in Q&A site and tags: the importance of prominent profiles in community productivity. *Journal of the Brazilian Computer Society*, 20(1), 1–16. DOI: https://doi.org/10.1186/1678-4804-20-5.

11. Gazan, R. (2007). Seekers, sloths and social reference: Homework questions submitted to a question-answering community. *New Review of Hypermedia and Multimedia*, 13(2), 239–248. DOI: https://doi.org/10.1080/13614560701711917.

12. Gazan, R. (2010). Microcollaborations in a social Q&A community. *Information processing & management*, 46(6), 693–702. DOI: https://doi.org/10.1016/j.ipm.2009.10.007.

13. Griffin, R. J., Neuwirth, K., Dunwoody, S., and Giese, J. (2004). Information sufficiency and risk communication. *Media Psychology*, 6(1), 23–61. DOI: https://doi.org/10.1207/s1532785x mep0601_2.

14. Hertzum, M., and Borlund, P. (2017). Music questions in social Q&A: an analysis of Yahoo! Answers. *Journal of Documentation*, 73(5) :992–1009. DOI: https://doi.org/10.1108/JD-02-2017-0024.

15. IResearch (2021), 2020 China Online Knowledge Questions and Answers Industry White Paper, Retrieved from https://pdf.dfcfw.com/pdf/H3_AP202008261401667082_1.pdf?1598459312000.pdf.

16. Jean, B. S. (2014). Devising and implementing a card-sorting technique for a longitudinal investigation of the information behavior of people with type 2 diabetes. *Library & Information Science Research*, 36(1), 16–26. DOI: https://doi.org/10.1016/j.lisr.2013.10.002.

17. Jin, J., Li, Y., Zhong, X., and Zhai, L. (2015). Why users contribute knowledge to online communities: An empirical study of an online social Q&A community. *Information & Management*, 52(7), 840–849. DOI: https://doi.org/10.1016/j.im.2015.07.005.

18. Kanayama, T. (2003). Ethnographic research on the experience of Japanese elderly people online. *New Media & Society*, 5(2), 267–288. DOI: https://doi.org/10.1177/146144480300500 2007.

19. Kayes, I., Kourtellis, N., Quercia, D., Iamnitchi, A., & Bonchi, F. (2015). The social world of content abusers in community question answering. *In Proc. the 24th international conference on world wide web*. ACM, New York, NY, 570–580. DOI: https://doi.org/10.1145/2736277.274 1684.

20. Kim, Y. C., Lim, J. Y., and Park, K. (2015). Effects of health literacy and social capital on health information behavior. *Journal of Health Communication*, 20(9), 1084–1094. DOI: https://doi.org/10.1080/10810730.2015.1018636.

21. Manafo, E., and Wong, S. (2012). Exploring older adults' health information seeking behaviors. *Journal of Nutrition Education and Behavior*, 44(1), 85–89. DOI: https://doi.org/10.1016/j.jneb.2011.05.018.

22. Oh, S., Zhang, Y. and Park, M. S. (2016). Cancer information seeking in social question and answer services: identifying health-related topics in cancer questions on Yahoo! Answers. *Information Research*, 21(3), 718. Retrieved from http://InformationR.net/ir/21-3/paper713.html.

23. Park, M. S., and Park, H. (2016). Topical network of breast cancer information in a Korean American online community: a semantic network analysis. *Information Research: An International Electronic Journal*, 21(4): 729. Retrieved from https://eric.ed.gov/?id=EJ1123250.

24. Pew Research Center (2021), Experts Say the "New Normal" in 2025 Will Be Far More Tech-Driven, Presenting More Big Challenges, Retrieved from https://www.pewresearch.org/internet/2021/02/18/experts-say-the-new-normal-in-2025-will-be-far-more-tech-driven-presenting-more-big-challenges/.

25. Pfeil, U., and Zaphiris, P. (2009). Investigating social network patterns within an empathic online community for older people. *Computers in Human Behavio*r, 25(5), 1139–1155. DOI: https://doi.org/10.1016/j.chb.2009.05.001.

26. Pieper, D., Jülich, F., Antoine, S. L., Bächle, C., Chernyak, N., Genz, J., and Icks, A. (2015). Studies analysing the need for health-related information in Germany-a systematic review. *BMC Health Services Research*, 15(1), 1–18. DOI: https://doi.org/10.1186/s12913-015-1076-9.

27. Shi, J., Li, C., Qian, Y., Zhou, L., and Zhang, B. (2019). Information Needs of Domestic and International HCQA Users——An Empirical Analysis. *Data Analysis and Knowledge Discovery*, 3(5), 1–10. https://doi.org/10.11925/infotech.2096-3467.2018.0813.

28. Sina news (2021), How big is the legacy of COVID-19? Some people haven't recovered for six months, Retrieved from https://news.sina.com.cn/c/2021-01-17/doc-ikftpnnx8533988.shtml.

29. United Nations News (2021), WHO: "Serial association" of symptoms after coronavirus will affect health care worldwide, Retrieved from https://news.un.org/zh/story/2021/02/1077892.

30. Vetsch, J., Fardell, J. E., Wakefield, C. E., Signorelli, C., Michel, G., McLoone, J. K., ... and ANZCHOG survivorship study group. (2017). "Forewarned and forearmed": Long-term childhood cancer survivors' and parents' information needs and implications for survivorship models of care. *Patient Education and Counseling*, 100(2), 355–363. DOI: https://doi.org/10.1016/j.pec.2016.09.013.

31. Viera, A. J., and Garrett, J. M. (2005). Understanding interobserver agreement: the kappa statistic. *Family medicine*, 37(5), 360–363. Retrieved from http://www1.cs.columbia.edu/~julia/courses/CS6998/Interrater_agreement.Kappa_statistic.pdf.

32. World Health Organization (2021), WHO Coronavirus Disease (COVID-19) Dashboard: Transmission Classification, Retrieved from https://covid19.who.int/info (accessed 8 July 2021).

33. Xiao, N., Sharman, R., Rao, H. R., and Upadhyaya, S. (2014). Factors influencing online health information search: An empirical analysis of a national cancer-related survey. Decision Support Systems, 57, 417–427. DOI: https://doi.org/10.1016/j.dss.2012.10.047.

34. Zhang, J., and Zhao, Y. (2013). A user term visualization analysis based on a social question and answer log. *Information processing & management*, 49(5), 1019–1048. DOI: https://doi.org/10.1016/j.ipm.2013.04.003.

35. Zhang, J., Chen, Y., Zhao, Y., Wolfram, D., and Ma, F. (2020). Public health and social media: A study of Zika virus-related posts on Yahoo! Answers. *Journal of the Association for Information Science and Technology*, 71(3), 282–299. DOI: https://doi.org/10.1002/asi.24245.

36. Zhao, W., Lu, P., Yu, S., and Lu, L. (2020). Consumer health information needs in China–a case study of depression based on a Social Q&A community. *BMC Medical Informatics and Decision Making*, 20(3), 1–9. DOI: https://doi.org/10.1186/s12911-020-1124-1.

Dan Wu, Ph.D., is a Professor in the School of Information Management at Wuhan University, a member of the Academic Committee of Wuhan University, and the director of the Human-Computer Interaction and User Behavior Research Center. Her research areas include information organization and retrieval, user information behavior, human-computer interaction, and digital libraries.

She has secured research grants and contracts, the National Social Science Foundation of China (Major Program), the National Natural Science Foundation of China (NSFC), and the Humanities and Social Sciences Foundation of the Ministry of Education.

Her work is published in various journals (including the IP&M, JASIST, etc.). She has presented at several conferences related to information behavior, information retrieval, and digital library (SIGIR, CIKM, CSCW, CHIIR, etc.).

Dr. Wu received her Ph.D. from the Peking University in Management. She serves as the Editor-in-Chief of Aslib Journal of Information Management and the Executive Editor of Data and Information Management. She also serves as a director at large of the ASIS&T. She is a member of the ACM Digital Library Committee, a member of the iSchool Data Science Curriculum Committee, and a member of the JCDL Steering Committee.

Le Ma is a Ph.D. candidate at the School of Information Management, Wuhan University. Her research interests include information retrieval, user information behavior, etc. She is a member of the Human-Computer Interaction and User Behavior Research Center of Wuhan University. She participated in several Chinese national and provincial research projects and published several academic papers during her doctoral study.

Digital Divide Faced by the Elderly Under the Covid-19 Pandemic

Shaobo Liang

Abstract

Under the Covid-19 situation, the older people's lack of understanding of the epidemic information, resulting in poor self-protection, limited travel, affects the purchase of life and medical items. These made a significant threat to their health. Due to the rules of home isolation, the Internet has become the primary source of information. But this leads to the digital divide faced by the elderly becoming more prominent, which has already existed before. This research studies the specific difficulties faced by the elderly in the digital divide in the epidemic. This paper used a semi-structured interview and recruited 25 older people to participate. By analyzing the interview content, this study used the content analysis method to analyze the difficulties of information acquisition and utilization of the elderly. In addition, this study analyzed the main reasons for the digital divide faced by the elderly from internal and external factors. Then, this study proposed suggestions for IT enterprises and government to solve the digital divide in the Covid-19 from the perspective of the elderly. Through this study, the findings can help solve the digital divide faced by the elderly in the epidemic, help them better obtain accurate information, and help them better protect their health.

Keywords

COVID-19 • Elderly • Information behavior • Digital divide

S. Liang (✉)
School of Information Management, Wuhan University, Wuhan, China
e-mail: liangshaobo@whu.edu.cn

Center for Studies of Human-Computer Interaction and User Behavior, Wuhan University, Wuhan, China

© The Author(s), under exclusive license to Springer Nature Switzerland AG 2023   75
X. Yuan et al. (eds.), *Social Vulnerability to COVID-19*, Synthesis Lectures
on Information Concepts, Retrieval, and Services,
https://doi.org/10.1007/978-3-031-06897-3_5

5.1   Introduction

In early 2020, the coronavirus disease-2019 (COVID-19) epidemic has spread rapidly to many countries worldwide and gradually became a global plague. Globally, as of 28 July 2021, there have been more than 195 million confirmed cases of COVID-19, including more than four million deaths, reported to WHO. This epidemic affected every aspect of our lives, especially global healthcare systems [21]. Shopping malls were forced to close, and many workers, especially women, lost their job. Trade volume between countries has decreased sharply, which seriously impacts the economy. Due to social inequality, some people face barriers to education, health care, employment, and access to information.

Vulnerable groups are more affected by the Covid-19. Older people seem to be more affected. Under the Covid-19 situation, the older people's lack of understanding of the epidemic information, resulting in poor self-protection, limited travel, affects the purchase of life and medical items. These made a great threat to their health. This epidemic has aroused the common concern of scholars in different fields around the world. In library and information science, Xie et al. [26] published an action proposal with other people from different countries, calling on scholars to pay attention to the information crisis under the epidemic.

Many countries use information technology to help solve the epidemic problem. But this leads to new problems for the elderly, such as the digital divide. In this information crisis [26], the challenges posed by the digital divide faced by the elderly seem to be more severe. With the development of digital socialization, the digital divide problem is becoming more and more obvious. The "digital divide" was originally used to describe the gap in access to new technologies between different populations [22]. In previous studies, the digital divide refers to the gap between people who benefit from the digital age and those who do not benefit from the digital age [19]. The digital divide also refers to the gap between people with different basic social informatization, digitization, and networking [8].

Those who cannot use the Internet and information technology will be weak in the digital society because they have difficulties accessing information, online shopping, and online learning, leading to unfair social phenomena. In the existing studies about the digital divide, researchers mainly focus on the difficulties of technology acceptance faced by people in some developing countries and the differences of the digital divide between different gender groups [2, 4].

However, the lives of ordinary people have been fully affected by the development of information technology. The elderly face more diverse new technologies and more technical challenges. Therefore, the situation of the digital divide has become more complex [11]. For instance, most older people access news information through traditional media such as television, radio, newspapers, etc. The timeliness of these methods is not strong, the reporting frequency is low, and the news content is single, making the elderly unable to understand the epidemic situation during the Covid-19 fully.

Therefore, this study aims to study the digital divide faced by the elderly in the context of the pandemic. The elderly respondents were recruited to participate in a semi-structured interview. This study focuses on the following research questions:

(1) What difficulties are caused by the digital divide faced by the elderly under the Covid-19 pandemic?
(2) What are the reasons for the digital divide among the elderly under the Covid-19 pandemic?
(3) What suggestions can help the elderly cope with the digital divide under the Covid-19 pandemic?

5.2 Related Works

With the progress of science and technology, especially the rapid development of the Internet, the digital divide has become an important problem faced by the elderly [14]. Solving the increasingly serious negative impact of the digital divide on the elderly has attracted attention in academia [23]. In the epidemic situation, a direct impact of the digital divide on the elderly is the access to health information. It is the focus of research in the field of information behavior. Information behavior research has always been a hot topic in library science and information science. In existing studies, scholars also pay attention to the information behavior of different populations, such as the elderly.

The health information behavior of the elderly has always been the focus of scholarly attention [8]. Although most people have widely accepted the Internet and smartphones, these technologies do not play their due role in making effective health information decisions for the elderly. This is considered to be the effect that health information literacy should achieve [7]. Ijiekhuamhen et al. [10] found that the primary information demand of the elderly is to solve health problems, the fairness of information resources, financial problems, and lack of information sources will affect the elderly's access to health information.

Chang et al. [1] proposed three core elements for the elderly to search for information: information search mode, reliability of health information, and decision-making route. They found that the elderly rely on social relationships to make health information choices. In terms of the influencing factors of the elderly's access to online health information, Rockmann and Gewald [18] investigated the impact of personality characteristics on the elderly's access to online health information. Liu et al. [15] believe that the health information literacy of the elderly includes the ability of information seeking, information understanding, and information utilization, and health information literacy can determine their health information behavior. Wu and Li [25] found that the health status of the elderly, their familiarity with the Internet, and the credibility of online health information are the main factors affecting the elderly to search online health information. Some scholars also pointed out [16] that older people with higher education and regular exercise have easier access to health information.

Since the epidemic, the elderly have faced various problems, such as social isolation, loneliness, and quality of life [12]. These problems are also related to the digital divide faced by the elderly [6]. Xie et al. [26] focused on the information crisis different populations faced when the epidemic broke out. Among them, the digital divide faced by the elderly during the epidemic is a prominent manifestation of the information crisis. Daoust [3] found that in the epidemic, the elderly are vulnerable groups, and the government does not pay enough attention to the information dilemma faced by the elderly. Sun et al. [20] also found that the current information dissemination channels make it more difficult for the elderly to obtain health information. Helping the elderly get more information related to the epidemic can effectively help the elderly protect themselves. Pan et al. [17] studied how vulnerable groups such as the elderly and children can better obtain information to solve the digital divide problem under the epidemic situation. The problem brought by the digital divide to the elderly is not only to obtain information but also to cause fear and death anxiety of the elderly [5, 13].

5.3 Research Design

5.3.1 User Interview

To answer the research questions, this paper designed a semi-structured interview to obtain the participants' answers. The interview was conducted in January 2020, when the epidemic broke out. The author published recruitment information on social media. Many respondents participate in this interview after their children see the information on social media.

Due to the epidemic's impact, everyone should abide by the regulations of community isolation and maintain social distance. Therefore, this interview was conducted via online meeting software. The interview time of each respondent is 25–35 min. The interview contents were recorded with a tape recorder, and the respondents were informed that all the interview contents were only used for scientific research.

5.3.2 Participants

We used a semi-structured interview and recruited 25 older people to participate. All participants were over 55 years old, and the oldest participant was 59 years old. All participants are from Wuhan, the Chinese city most seriously affected by the epidemic. These participants felt the most about the digital divide they faced during the epidemic. After the interview, each participant was paid $10.

5.3.3   Data Analysis

After obtaining the interview records of all participants through the recorder, a machine transcription platform[1] was used to convert the recording into text. Then, the author recruited researchers to proofread the transcribed text to ensure accuracy. Through transcribing, this paper has obtained 13,235 words.

This paper used the content analysis method to analyze the interview content. The qualitative analysis software MAXQDA was used to code the interview content. This paper recruited two Ph.D. candidates familiar with content analysis to encode the interview content. The coding process mainly includes primary coding and secondary coding. The coder first read all the interview contents sentence by sentence and established the first-level coding. Based on first-level coding, coders carried out secondary coding and summarized the relationship between first-level coding categories.

After coding, Cohen's kappa coefficient test was performed on the coded documents. After the reliability test of the coding results, the average Cohen's kappa coefficient is 0.846. If the coder's coding results of the sentences in the interview content are inconsistent, the author and the two coders will discuss them together to form the final coding.

5.4   Findings

5.4.1   Difficulties Caused by the Digital Divide Faced by the Elderly

Due to the low Internet and smartphone usage, the elderly may face difficulties such as being unable to take the bus, shopping in supermarkets, registering online in hospitals, and doing business online. Their daily life is deeply affected. For example, many older people do not know the rules of home isolation. Some older people even believe that drinking disinfectants can keep them healthy and defend Covid-19.

This paper first analyzes the difficulties caused by the digital divide faced by the elderly in the epidemic. According to the data analysis method in Sect. 5.3.3, this paper found that the difficulties caused by the digital divide mainly include two aspects, as shown in Table 5.1.

The first major difficulty is evaluating information authenticity, including evaluation of rumors on social media and fake news. For example, the participant said, *"Some people in WeChat release rumors, but I can't judge the truth"*. During the epidemic period, social media was filled with a lot of false information due to various reasons, which caused trouble to the elderly who rarely surf the Internet. A participant said, *"There is a lot of news about the epidemic. I can't tell the true from the false"*.

---

[1] https://global.xfyun.cn/.

Table 5.1 Coding results on difficulties caused by the digital divide faced by the elderly

Level 2 coding results	Level 1 coding results		Frequency
	Code	Example	
Evaluation of information authenticity	Evaluation of rumors on social media	*"Some people in WeChat release rumors, but I can't judge the truth." etc.*	17
	Evaluation of fake news	*"There is a lot of news about the epidemic. I can't tell the truth from the false." etc.*	14
Access to digital services	Use of health code	*"I can't use the health code, which brings me trouble," etc.*	19
	Online shopping	*"During the epidemic, I can't go shopping, but I'm not proficient in online shopping." etc.*	16
	Electronic medical treatment	*"I'm not feeling well and can't go to the hospital. But I don't know how to ask online." etc.*	12
	Mobile finance	*"I want to withdraw money, but I can't go out. I don't have a mobile banking app." etc.*	9

The second major difficulty is access to digital services, including health codes usage, online shopping, electronic medical treatment, and mobile finance. The most important ones are electronic medical treatment and mobile finance.

According to the level-1 coding results in Table 5.1, the main problem caused by the digital divide is the access to digital services, followed by the evaluation of information authenticity. These two difficulties involve two important links in information behavior: information acquisition and information utilization.

5.4.2 Reasons for the Digital Divide

This part mainly discusses the reasons for the digital divide among the elderly under the Covid-19 pandemic. This study mainly analyzed the factors leading to the digital divide from the perspectives of personal factors and external factors through interviews. By coding the interview content, personal factors mainly include physical function degeneration, poor learning ability, and psychological problems. External factors include low Internet usage and changes in means of media communication.

5.4.2.1 Personal Factors

Personal factors refer to factors related to users' physiology and psychology. Among personal factors, the main cause of the digital divide among the elderly is poor learning ability. The code "Information system interaction" means that there are obstacles in the interaction between the elderly and the information system. For example, a participant said, "*My mobile phone operation is very complicated*". The code "New APP adoption" refers to the obstacles when the elderly use some new apps, mainly reflected in the perceived ease of use. In the epidemic environment, some older people have to download some new APPs, such as newspaper APP, to obtain information related to the epidemic. However, these older people can't use the new APP smoothly, leading to the digital divide problem.

Usually, young people are curious about new information technology and have a strong ability to accept new information technology and higher efficiency. In contrast, the ability of older people to accept information technology is low. Through Table 5.2, "psychological problems" are also an important factor, which is rarely mentioned in previous studies. When the elderly live with their children, they will be shy to ask questions related to new technology. This will make them feel embarrassed, resulting in a certain amount of social pressure. This is also a very interesting factor.

5.4.2.2 External Factors

External factors correspond to personal factors and refer to environmental reasons. This paper mainly identifies two external factors through the coding of the interview content.

Table 5.2 Coding results on personal factors

Level 2 coding results	Level 1 coding results		Frequency
	Code	Example of interview contents	
Physical function degeneration	Visual impairment	"*I can't read the text on my mobile phone,*" etc.	16
	Touch impairment	"*Sometimes I touch my cell phone and don't respond,*" etc.	11
Poor learning ability	Information system interaction	"*My mobile phone operation is very complicated,*" etc.	22
	New APP adoption	"*I installed an app, but it is very inconvenient to use,*" etc.	18
Psychological problems	Social Pressure	"*I was embarrassed to ask young people to operate my mobile phone.*" etc.	18

Table 5.3 Coding results on external factors

Level 2 coding results	Level 1 coding results		Frequency
	Code	Example of interview contents	
Low usage of internet	Low usage of mobile Internet	*"I seldom surf the Internet using a mobile phone and don't know where to get information related to the epidemic," etc.*	13
	Never used the Internet	*"I don't have a smartphone, and I can't surf the Internet," etc.*	4
Changes in means of media communication	The Internet has become the main channel of news	*"Now young people use their mobile phones to watch the news. We seldom use it, so it's slow to know the news," etc.*	18

According to Table 5.3, among external factors, the main cause of the digital divide is low Internet usage, including "Low usage of mobile Internet" and "Never used the Internet". Especially affected by the changes in means of media communication, the elderly have fewer channels to obtain information than the young, which leads to the difficulties of the elderly in obtaining information. As the participant said, *"Now young people use their mobile phones to watch the news. We seldom use it, so it's slow to know the news".*

5.4.3 Suggestions to Solve the Digital Divide

In the above sections, this paper discusses the problems caused by the digital divide faced by the elderly in the epidemic and the causes. This part answers the third research question: suggestions can help the elderly cope with the digital divide under the Covid-19 pandemic.

Through the interview content analysis, the participants mainly put forward suggestions to the two subjects, hoping they could do more to help the elderly solve the digital divide, as shown in Table 5.4.

For IT enterprises, what they can do is mainly the design of Internet products. Firstly, Internet companies need to provide more optimized interface design and interaction design for the elderly. These optimized designs can help the elderly use new products or APPs more conveniently and quickly. In addition, Internet companies need to pay attention to the elderly when providing guidelines for new users, such as designing animation and voice, to help the elderly learn to use the new app more easily.

Table 5.4 Coding results on suggestions to solve the digital divide

Level 2 coding results	Level 1 coding results		Frequency
	Code	Example of interview contents	
IT enterprise	More optimized design	*"Internet companies can pay more attention to the use habits of the elderly in the interface design of the app." etc.*	13
	Guidance for the elderly	*"When using a new app, it can provide us with more guidance, such as voice guidance." etc.*	7
Government	More investment	*"I think the government should invest more resources to help us learn information technology and the Internet," etc.*	9
	Support provided by the public service	*"Hospitals and public transportation cannot only be paid electronically. We should take into account that our elderly can't use mobile phones" etc.*	17

Participants also believed that the government should also help the elderly solve the digital divide. Public service institutions should help the elderly solve technical problems, especially those who have difficulties using information technology. For example, the hospital should help the elderly without online appointment to get services quickly. The railway station should help the elderly without health codes to travel normally. It is worth noting that scientific research has always been an effective means to solve the digital divide, from technology, policy, etc. But for ordinary people, scientific research is not their focus. What they want most is support from public services.

5.5 Discussions

5.5.1 Reasons for the Digital Divide in Covid-19

This paper summarizes the main reasons for the digital divide under the epidemic situation through the interview content analysis.

There are barriers for the elderly to learn to use the Internet. The first is the personal factor. Most older people's physical function degenerates, and their learning ability is poor. The Internet interaction process is complex for them, the font of mobile devices is small, and it is inconvenient to use. At the same time, the elderly are more likely to be affected by the concept, living habits, and education level and tend to use traditional media.

The second is external factors. First of all, the frequency of the elderly using the mobile Internet is too low. According to the forty-fifth report of CNNIC, the number of Internet users in China is 904 million, but only 6.7% of them are 60 years old or above. The Internet has become the main channel of news communication. Online news has the characteristics of freshness and timeliness, which means that the higher the frequency of Internet access, the higher the possibility of receiving the latest news. So, the audience relying on traditional media may miss the latest news. Faced with the covid-19 and other emergency events, missing information greatly impacts the elderly's daily lives, such as clothing, food, and travel. For example, due to the lack of information, older people worldwide generally faced difficulties buying food during the Covid-19.

5.5.2 How to Solve the Digital Divide

To better help the vulnerable groups (such as the elderly) in the covid-19 or other emergency events. We think we should solve this problem from the following aspects.

IT enterprise can provide more convenience for the elderly in technical means. In order to encourage the elderly to use the network, they should optimize the design of mobile devices. At the same time, the elderly have a weakness in hearing, vision, and memory. At present, most mobile phone APPs focus on young and middle-aged groups, and there is a lack of consideration for elderly users in product design. In addition to the elderly's reasons, the non-applicability of digital products also increases the burden of the elderly. For example, participants complained: "*a few years ago, my son bought an Internet TV for his family. But the operation interface is too complex, so it's hard to find a program you want to watch*".

At present, the digital development of some developing countries has not reached the fully developed stage. It is still in gradual transition from traditional society to digital society. In order to develop rapidly, some institutions and businesses blindly pursue innovation and change in design concepts and product functions without taking into account the habits of elderly users. In fact, the "marginalization" of the elderly by digital life is also a great loss of the digital economy. There is still much room to tap the online consumption potential of the elderly. Internet enterprises should develop more products and services that are safe to use, simple to operate, and in line with the living habits of the elderly.

For example, the designers of smart phones for the elderly need to simplify Internet operation. All kinds of news APPs can use simple pages and large fonts for the elderly. The smartphone should actively push the topics of their concern, such as health, weather, convenient services. For example, Alipay established Alipay University for the elderly through online and offline courses to help the elderly break away from the electronic barriers of mobile phones. In addition, Alipay also introduced to the elderly special Alipay applet, which includes scanning, payment, and other common functions, and intimate to enlarge the font, easy to use.

The government should also focus on the elderly in rural areas, especially in developing countries. Many older people in rural areas do not have smart phones. Although some older people have smartphones, it isn't easy to operate. Some older people can't even input words because of their low level of education.

In addition, although the participants did not mention how to solve the problem of fake news, this study believes that it is very important for the elderly to eliminate false network information, especially false health information.

5.6   Conclusion

This research studies the difficulties faced by the elderly in the digital divide during the Covid-19. A semi-structured interview was conducted, and 25 older people were recruited to participate. This paper used the content analysis method to analyze the interview content and identified the difficulties of information acquisition and utilization of the elderly caused by the digital divide. In addition, this study analyzed the main reasons for the digital divide faced by the elderly from internal and external factors. Then, this study proposed suggestions for IT enterprises and government to solve the digital divide in the Covid-19. Through this study, our findings can help solve the digital divide faced by the elderly in the epidemic, help them better obtain accurate information, and help them better protect their health.

Acknowledgements This work is supported by the National Natural Science Foundation of China (No. 72104187) and is also supported by China Postdoctoral Science Foundation (No: 2021M692480).

References

1. Chang, L., Basnyat, I., & Teo, D. (2014). Seeking and processing information for health decisions among elderly Chinese Singaporean women. Journal of Women & Aging, 26(3), 257–279.
2. Cooper, J. (2006). The digital divide: The special case of gender. Journal of Computer Assisted Learning, 22(5), 320–334.
3. Daoust, J. F. (2020). Elderly people and responses to COVID-19 in 27 Countries. PloS One, 15(7), e0235590.
4. Di Maggio, P., & Hargittai, E. (2001). From the 'digital divide'to 'digital inequality': Studying Internet use as penetration increases. Princeton: Center for Arts and Cultural Policy Studies, Woodrow Wilson School, Princeton University, 4(1), 4–2.
5. Dreisiebner, S., März, S., & Mandl, T. (2020). Information behavior during the Covid-19 crisis in German-speaking countries. arXiv preprint arXiv:2007.13833.

6. García-Fernández, L., Romero-Ferreiro, V., López-Roldán, P. D., Padilla, S., & Rodriguez-Jimenez, R. (2020). Mental health in elderly Spanish people in times of COVID-19 outbreak. The American Journal of Geriatric Psychiatry, 28(10), 1040–1045.

7. Hallows, K. M. (2013). Health Information Literacy and the Elderly: Has the Internet Had an Impact? Edited by Rick Block. The Serials Librarian, 65(1), 39–55.

8. Hilbert, M. (2011). The end justifies the definition: The manifold outlooks on the digital divide and their practical usefulness for policy-making. Telecommunications Policy, 35(8), 715–736.

9. Hu, X., Wang, J., & Wang, L. (2013). Understanding the travel behavior of elderly people in the developing country: a case study of Changchun, China. Procedia-Social and Behavioral Sciences, 96, 873–880.

10. Ijiekhuamhen, O. P., Edewor, N., Emeka-Ukwu, U., & Egreajena, D. E. (2016). Elderly people and their information needs. Library Philosophy and Practice.

11. Jackson, L. A., Zhao, Y., Kolenic III, A., Fitzgerald, H. E., Harold, R., & Von Eye, A. (2008). Race, gender, and information technology use: The new digital divide. CyberPsychology & Behavior, 11(4), 437–442.

12. Kasar, K. S., & Karaman, E. (2021). Life in lockdown: Social Isolation, Loneliness and Quality of Life in the Elderly During the COVİD-19 Pandemic: A Scoping Review. Geriatric Nursing.

13. Khademi, F., Moayedi, S., Golitaleb, M., & Karbalaie, N. (2020). The COVID-19 pandemic and death anxiety in the elderly. International Journal of Mental Health Nursing, https://doi.org/10.1111/inm.12824. Advance online publication. https://doi.org/10.1111/inm.12824.

14. Kiel, J. M. (2005). The digital divide: Internet and e-mail use by the elderly. Medical Informatics and the Internet in Medicine, 30(1), 19–23.

15. Liu, Y. B., Liu, L., Li, Y. F., & Chen, Y. L. (2015). Relationship between health literacy, health-related behaviors and health status: A survey of elderly Chinese. International Journal of Environmental Research and Public Health, 12(8), 9714–9725.

16. Oh, Y. S., Choi, E. Y., & Kim, Y. S. (2018). Predictors of smartphone uses for health information seeking in the Korean elderly. Social Work in Public Health, 33(1), 43–54.

17. Pan, S. L., Cui, M., & Qian, J. (2020). Information resource orchestration during the COVID-19 pandemic: A study of community lockdowns in China. International Journal of Information Management, 54, 102143.

18. Rockmann, R., & Gewald, H. (2015). Elderly People in eHealth: Who are they?. Procedia Computer Science, 63, 505–510.

19. Smith, C. W. (2002). Digital corporate citizenship: The business response to the digital divide. Indianapolis: The Center on Philanthropy at Indiana University. ISBN 1884354203.

20. Sun, Z., Yang, B., Zhang, R., & Cheng, X. (2020). Influencing factors of understanding COVID-19 risks and coping behaviors among the elderly population. International Journal of Environmental Research and Public Health, 17(16), 5889.

21. Usher, K., Bhullar, N., Durkin, J., Gyamfi, N., & Jackson, D. (2020). Family violence and COVID-19: Increased vulnerability and reduced options for support. International Journal of Mental Health Nursing, 29(4), 549–552. https://doi.org/10.1111/inm.12735.

22. Van Dijk, J. A. (2006). Digital divide research, achievements and shortcomings. Poetics, 34(4–5), 221–235.

23. Van Jaarsveld, G. M. (2020). The effects of COVID-19 among the elderly population: a case for closing the digital divide. Frontiers in Psychiatry, 11.

24. WHO. WHO Coronavirus (COVID-19) Dashboard. https://covid19.who.int/.
25. Wu, D., & Li, Y. (2016). Online health information seeking behaviors among Chinese elderly. Library & Information Science Research, 38(3), 272–279.
26. Xie, B., He, D., Mercer, T., Wang, Y., Wu, D., Fleischmann, K. R., ... & Lee, M. K. (2020). Global health crises are also information crises: A call to action. Journal of the Association for Information Science and Technology, 71(12), 1419–1423.

Shaobo Liang, Ph.D., is an Assistant Professor in the School of Information Management at Wuhan University, Wuhan, China. His research interests lie in understanding user information-seeking behavior, mobile search behavior, and cross-device search.

He has secured research grants and contracts, including the National Natural Science Foundation of China (NSFC) and the China Postdoctoral Science Foundation. His work is published in various journals, and he has presented at several conferences related to information retrieval, digital society, and user behavior.

Dr. Liang received his Ph.D. from Wuhan University in Management Science. He is an officer board for the ASIS&T Asia-Pacific Chapter and serves as a webmaster and ACM member. He serves for several conferences, including ASIS&T, JCDL, ICADL, as a PC member and reviewer.

Difficulties of Vulnerable Groups in Accessing the Public Information Service During COVID-19 Pandemic in China

6

Kun Huang, Lei Li, Shi Chao Luo, and Xiao Yu Wang

Abstract

The outbreak of COVID-19 pandemic seriously threatens people's lives and health all over the world. Public information service is an important way for the public to understand the pandemic development and pandemic prevention and control measures. However, more people become vulnerable rather than the typical vulnerable groups due to the pandemic encounter difficulties in accessing public information services. To better help vulnerable people during the emergencies, based on related vulnerable groups theory, this chapter identifies two types of vulnerable groups in China. It collects information from news, journal papers, conference papers and other relevant perspectives to examine the difficulties that vulnerable people have encountered. Moreover, this chapter puts forward some suggestions from the aspect of policies and regulations, technologies and information systems, service content, and operating mechanism of public information services to better meet the information needs of vulnerable people.

Keywords

Vulnerable groups • Difficulties • Public information service • Pandemic

6.1   Introduction

In early 2020, the outbreak of COVID-19 pandemic seriously threatened people's lives and health and affected the order of production and life in different countries all over

K. Huang · L. Li (✉) · S. C. Luo · X. Y. Wang
School of Government, Beijing Normal University, Beijing, China
e-mail: leili@bnu.edu.cn

© The Author(s), under exclusive license to Springer Nature Switzerland AG 2023
X. Yuan et al. (eds.), *Social Vulnerability to COVID-19*, Synthesis Lectures
on Information Concepts, Retrieval, and Services,
https://doi.org/10.1007/978-3-031-06897-3_6

the world. As reported by the World Health Organization, by July 18, 2021, there have been 188,655,968 confirmed cases of COVID-19, including 4,067,517 deaths [46]. By the end of 2020, the epidemic caused 114 million people to lose their jobs, 81 million people out of the labor force [2]. And the global economy was expected to lose over $22 trillion in the future 5 years since 2020 [10]. Because of the pandemic, the traditional vulnerable groups, who are relatively disadvantaged in the aspects of economy, culture, physical fitness, intelligence, social situation, lack of resources, poor economic quality, low quality of life and vulnerable groups and ranks, suffer from more severe difficulties [12]. Moreover, the high risk of virus infection, epidemic prevention and control policies and measures, may lead to more people becoming vulnerable groups. In the COVID-19 pandemic, vulnerable groups are not only the elder, those with ill health and comorbidities, or homeless or underhoused people, but also people from different social and economic classes that might struggle to cope financially, mentally, or physically with the crisis [21]. To fight the pandemic, public information service (PIS) plays an important role, which helps people understand the epidemic status, anti-epidemic measures and assists people in making scientific and reasonable decisions. However, infodemic caused by a large increase of misinformation in a short period of time makes it difficult for the public to obtain reliable, authentic and authoritative public information [63]. Moreover, information technologies employed in the epidemic prevention and control measures make both typical vulnerable groups and ordinary people face more challenges and problems in accessing and using PIS, which directly affects their living conditions. Thus, insight and understanding of difficulties of vulnerable groups in accessing public information during the emergencies, and investigation of the bottleneck of PIS in meeting the information needs of vulnerable groups have positive practical significance for effectively promoting the PIS, improving the social rights and interests of vulnerable groups, and forming a unified epidemic prevention consensus and action of the whole society. Since the epidemic situation is different in different countries, the anti-epidemic policies and measures adopted by the governments are also different. Therefore, taking China as an example, this study focuses on the following issues:

- What difficulties do the vulnerable groups in China encounter when trying to access PIS?
- How to improve the PIS to help vulnerable groups survive in public health emergencies?

6.2 The New Definition of Vulnerable Groups in COVID-19 Pandemic

Vulnerable groups are always vulnerable in all aspects of social life, such as politics, economy, culture, and physical health. Typical vulnerable groups usually include women, children, the elderly, the disabled, ethnic minorities or ethnic minorities. Due to the impact of the COVID-19 pandemic, people who are not vulnerable may become vulnerable. Chang [4] defines vulnerable groups from two dimensions. The first dimension is the impact of epidemic situations and epidemic prevention and control measures on people,The second dimension is the influence of the subject's ability and situation. Based on Chang's work, two types of vulnerable groups during the pandemic could be identified.

6.2.1 Category I

The first type of vulnerable groups are those who lack the corresponding ability and are threatened by the epidemic because of their physical conditions or prevention measures. These vulnerable groups include the typical vulnerable group like visually impaired, hearing impaired, mental disabled, the elderly, children, pregnant women, as well as the confirmed cases of COVID-19, patients with serious preexisting disease, and those who are stranded in other places due to epidemic prevention and control, restricted range of activities and unnecessary commercial circulation.

On the one hand, public health emergencies further aggravated the living difficulties of typical vulnerable groups like the elderly living alone and the disabled who need care. The Chinese Center for Disease Control and Prevention has published a large case series to report the pandemic situation in China [52]. It demonstrated that among 44,672 confirmed cases of COVID-19 (updated through 11 Feb 2020), 87% of the patients were 30–79 years. The overall case-fatality rate (CFR) was 2.3%, while cases in those aged 70–79 had an 8.0% CFR and cases in those aged 80 years and older had a 14.8% CFR. CFR was elevated among those with preexisting comorbid conditions—10.5% for cardiovascular disease, 7.3% for diabetes, 6.3% for chronic respiratory disease, 6.0% for hypertension, and 5.6% for cancer. Moreover, when epidemics intensify, people with mental health disorders are generally more susceptible to infections [60]. It was reported that 323 severe mental disorders were confirmed COVID-19 by February 18, 2020, and 43 cases were suspected of having new crown pneumonia, which covered 17 provinces in China National Health Commission of the PRC [31].

On the other hand, some people were stranded out of town because of epidemic prevention and control measures, resulting in no income and lack of shelter and food [66]. As of February 29, 2020, a total of 1573 people living in difficulties were rescued in Wuhan. The amount of aid invested was over 3 million yuan [45]. From March 25 to April 8, 2020, more than 61,000 residents of Beijing who stayed in Hubei have returned to Beijing safely [51]. In addition, hospital staff, hospital wards, drugs, and supplies were urgently used for epidemic prevention and control due to the need to treat patients with COVID-19 [9], Office of the CPC Ministry of Industry and Information Technology of the PRC, [35]. As a result, the medical resources that can provide other patients were greatly reduced [15]. Similarly, confirmed cases crowded the emergency rooms in every hospital in the winter of 2020, during the COVID-19 pandemic in the USA [34]. The oxygen bottles were almost used up. Cancer patients and other severe patients were hard to find beds. Some had to go home to take conservative treatment or even to stop treatment [1]. A large number of asymptomatic patients furtherly increased the pressure of hospital reception. Many hospitals had to suspend routine medical activities, which aggravated the survival difficulties of patients with preexisting severe diseases and major diseases.

6.2.2 Category II

The second type of vulnerable groups are those in a disadvantageous environment, which makes them vulnerable to the threat of virus, or lose their financial resources because of prevention and control measures. These vulnerable groups can be divided into four subcategories. The first are the front-line health care workers, close contacts of infected people. The second are the courier, takeaway, public transport drivers who face a high risk of infection. The third are those in the care state, such as patients in mental hospitals, the elderly in nursing institutions, and prisoners in prisons. The fourth are those who have lost their jobs or have no income due to the suspension of business, shops and projects caused by the epidemic prevention and control measures.

As of 31 December 2020, the International Council of Nurses (ICN) data set revealed [17] that over 1.6 million healthcare workers had been infected in 34 countries. And, the ICN believes that around 10% of all confirmed COVID-19 infections are among healthcare workers, with a range of 0–15%. In addition, COVID-19 is more transmissible than SARS and MERS in households, and it can be transmitted even in the viral incubation period [19]. Family transmission is one of the main ways of transmission of COVID-19. According to relevant studies, once the first person in a previously healthy family gets infected with COVID-19, the others have about a 17% chance of getting infected [14]. The close contacts of infected people like family members, colleagues in the same office or neighbors in the same community are all vulnerable groups and temporarily be restricted to travel and accept epidemiological investigation.

6.3 The Difficulties of Vulnerable Groups in Accessing PIS

6.3.1 Methodology

To collect the difficulties of access to PIS for vulnerable groups during the COVID-19 pandemic, we took China as an example and gathered information from news, journal papers, conference papers and other relevant perspectives. We used several keywords to search related information, including "COVID-19", "vulnerable groups", "disadvantaged groups", "vulnerable populations", "deaf", "blind", "handicapped", "courier", "elderly", "children", "patient", "quarantined people", "medical staff", "unemployed", "information service", "public information" and "information seeking" et al. We searched in databases such as CNKI, WanFang, PubMed, ScienceDirect, Wiley Online Library, Web of Science, and LISTA. Moreover, we collected news through Baidu, the most popular search engine in China, and news portals, including People's Daily Online, Xinhua Net, Sina.com etc. Based on the news reports and literature, we identified the difficulties of those two vulnerable groups.

6.3.2 Category I

People with disabilities are typical vulnerable groups. For the profoundly deaf, information about the epidemic situation can be obtained through television, live network news conference and other means. However, if the news and live broadcast are not equipped with sign language translation or only broadcast headlines or keyword letters, it is very difficult for deaf people to understand the whole contents of the press conference [40]. According to the joint survey data released by the former National Health and Family Planning Commission and China Disabled Persons' Federation in 2016, the number of people with hearing impairment in China was 206 million, accounting for 15.84% of the total population [36]. Although most countries have sign language translation when they broadcast major news events, the special and urgent features of the COVID-19 epidemic seriously affect the working conditions and efficiency of government agencies and departments, resulting in the lack of timely sign language personnel or considering the information acquisition needs of the disabled. The White House was also sued by the National Association of the Deaf for not having a sign language translator at the daily epidemic briefing [38]. In Feb 2020, The Beijing Deaf Association launched an initiative through the Beijing Daily suggesting that press conferences on the situation of the epidemic could have sign language translation or subtitles [40].

 For people with visual impairment, the difficulty of information use mainly focuses on visual information reading. To make PIS related to pandemic more efficient and intuitive, many new technologies have been applied, such as the real-time epidemic map based on big data and visualization technology, which presents the risk level of different regions in different colors and provides rich interactive ways. However, the screen reading function of smartphones can only read part of the map information, and it is difficult for people with visual impairment to fully understand the epidemic information and popular science information presented by visual means such as pictures [65]. In addition, the application of heath kit also brings many difficulties to the disabled. For example, in Beijing, Mr. Li, a blind man, failed to complete registration for a long time after the implementation of the "Beijing Health kit ", because he couldn't complete face recognition alone. Although Mr. Li used the voice assistant function, face recognition requires high operation accuracy. The round frame of face recognition in "Beijing health kit" was a bit small, only the size of a dollar coin. If the angle between the face and the round frame was incorrect, the distance was too far or too close, it will fail to complete recognition successfully. Li gave up after trying many times [61]. Besides, many information service platforms used the graphic verification code during user login. But screen reading software only prompts that this is a picture and cannot read the text embedded in the picture, so the visually impaired cannot complete login verification alone [61].

Epidemic prevention measures also make it difficult for people with hearing and visual impairment to access information. For example, "lip talk" is a commonly used auxiliary means for face-to-face communication among deaf people. For the need of epidemic prevention and control, people are required to wear masks in public areas to block people's noses and mouths, which makes it more difficult for deaf people to communicate with others, especially community social workers who carry out prevention and control work [3].

For the elderly, their vision, hearing, other senses, and learning ability have declined due to aging. The senior people's ability to accept and learn new technologies becomes limited, and most of them are not good at using smartphones, computers and other information technology software and tools. For example, many old people don't know how to use the "Health Kit", which makes daily travel difficult [70]. In the early stage of the pandemic, masks, alcohol and other scarce epidemic prevention materials were mostly obtained through official websites and mobile app. The opening hours and flow restrictions of some restaurants and parks were also mostly released online, which greatly restricted the rights and interests of the elderly to obtain such public services [67]. In addition, epidemic prevention measures limit offline communication. Online communication and circles of acquaintances have become the primary information channels for the elderly. Since the elderly always think that they have rich life experiences, they may be too confident to believe rumors [44].

For children, they are not mature in physiology, cognition, emotion, social and other aspects, and are more vulnerable than other groups in public health emergencies [42]. Their information mainly comes from the Internet, smart devices and parents, but the amount of information easy for children to understand is small [62]. In addition, 56.38% of urban children hardly went out during the pandemic, two or three generations have been living in an indoor environment, which could be quite challenging for children and parents and might increase the probability of psychological problems in children [13]. After the students return to school, fierce competition, the parents' neglect of the students' learning pressure and the children's weak resilience led to some students' excessive pressure and suicidal tendency [55].

For the patients suffering from COVID-19, the difficulties in getting needed information, lack of trusted access to information and not knowing what search terms to use are the main difficulties in information seeking. In addition, poor identification ability, lack of relevant domain knowledge and lack of opportunities for discussion are the main barriers to health information use. Patients over 60 years old lacked health knowledge and information literacy, which resulted in low demand and use of health information. They also avoid psychologically relevant and bias-related information [47].

For the people stranded because of the epidemic prevention and control policy, how to solve the problem of basic living was their big concern. In the first month of Wuhan's closure, some low-income groups stranded in Wuhan had to live temporarily in garages, parks, underground passages and hospitals. On February 25, 2020, the epidemic prevention and control headquarters of Hubei Province issued a circular, requiring local governments and relevant parties to provide rescue services for all kinds of people staying in Hubei Province, and provide basic living security such as accommodation and medical care [24]. In April, 2020, local civil affairs departments opened up 864 temporary relief places [8]. However, there were still difficulties for vulnerable groups to receive and understand the policies. Mr. Huang, who lived in Xiaogan, Hubei Province, worked as a porter in Wuhan. Because of the epidemic, he was unable to return home and couldn't afford to rent a house. Although he read the news, he never noticed the relief documents issued by Wuhan civil affairs department. With the occasional help of a volunteer, Mr. Huang learned about the policy and logged on to the official website of the Wuhan Civil Affairs Bureau to submit applications [49].

For the quarantined people in the outbreak area, both the elderly living alone and the disabled who lacked personal care were facing many difficulties in life. On February 17, 2020, the most stringent 24-h closure measures were implemented in all villages, communities, communities and residential areas in Hubei Province [16]. Although the community provided residents with basic living security, residents still felt that they lacked information, they always received delayed information and were cheated by false information when seeking information about living materials [5, 39, 72].

For patients suffering from malignant tumors, uremia, hypertension, diabetes, and other diseases, the pandemic has put them in greater difficulties. These patients needed to go to the hospital regularly for examination or treatment. However, patients of COVID-19 had increased dramatically during the outbreak, and the medical resources for special patients had been reduced significantly [15]. Patients had to turn to online medical platforms and social media for help due to the lack of hospital reception ability [6, 29]. Although the government also carried out voluntary assistance activities through social platforms, such as microblog and Chaohua "help for non-pneumonia patients" [22], the information on the Internet and social platforms was unclear and confusing, which made it more difficult for patients to obtain authoritative and useful information effectively.

6.3.3 Category II

Front-line medical staffs face the highest exposure risk. Their main problems are virus prevention, mental health and sleep disorder. During the epidemic, medical staff paid close attention to the information such as clinical characteristics of virus infection, epidemic characteristics of diseases, and disposal specifications [56]. Some young medical staff with insufficient experience and weak professional ability were vulnerable to the impact of the information epidemic and believed in rumors, [25]. In addition, the psychological sub-health problems and moderate/severe psychological problems of front-line medical staff had also received much attention [20]. Information avoidance was found for the group of clinical personnel who had serious psychological problems. They were eager for personalized one-to-one counseling as a treatment option, but only 21% of medical staff with severe mental disorders had received individual or group psychological counseling services [20].

For the delivery workers, during the pandemic, they needed to travel around various streets, stores, and communities, facing the risk of infection by the face-to-face communication and the goods or takeaways contaminated by the virus [54]. Moreover, access control policies varied in different communities. As Time Weekly (2020) reported, these delivery drivers were risking their health to keep China running during the pandemic. In addition, the profits of the logistics industry decreased due to the pandemic, which also led to various problems of overwork and improper salary reduction. Most of the delivery workers are junior and senior high school educated [27], they lack awareness of labor rights. Thus, they don't know where to get information about rights protection [48]. When labor rights are violated, most delivery workers choose to talk to their friends or resign, rather than ask for help from units or relevant departments [68].

Prisoners serving their sentences, suspects in custody, and prison guards become vulnerable during the pandemic. The limited channel of information and the lack of attention to information disclosure make it easy for misinformation to spread in the prison and custody. In addition, most prisoners have low education level, poor information literacy, lack of judgment and are easy to believe misinformation [59]. In April 2020, a rumor of an epidemic outbreak in Changsha Women's Prison in Hunan Province spread widely, and the official refuted the rumor promptly [33]. It was fortunate that the official clarified the rumors in time, or it would cause panic among the prisoners and have serious consequences.

For patients who have severe mental disorders in psychiatric hospitals, they often have cognitive impairment and can't accurately understand pandemic information [58], mainly relying on supervision and nursing of medical staff. In economically underdeveloped areas, public health education also needs to be strengthened in some psychiatric hospitals. A survey on the preferred COVID-19 information sources for inpatients of five psychiatric hospitals in Gansu Province indicated that one-third of patients obtained information from medical staff [69]. However, the COVID-19 training program for medical staff didn't fully cover everyone in psychiatric hospitals. Only 64.6% of mental health professionals received relevant training in the early stage of the COVID-19 outbreak in Chinese psychiatric hospitals, according to a study of knowledge and attitudes of medical staff in two Chinese psychiatric hospitals regarding COVID-19 [41], which would greatly affect the health and life safety of mental patients.

For the unemployed, re-employment had become their greatest concern. To deal with the COVID-19 virus and reduce its transmission, many countries have taken measures to stop production. Non-essential public places were closed. Shops, cinemas, and restaurants were out of business. That caused that many people to temporarily lose their sources of income. Furthermore, enterprises were also under tremendous pressure because of the suspension of production. Some small and medium-sized enterprises were affected and plunged into the financial crisis [71]. Even though the government has taken some assurance measures, the risk of infringement of ordinary workers' legitimate rights and interests was still high [11]. Some workers were dismissed without compensation, or their severance packages were cut for various reasons [50]. On the one hand, ordinary workers usually lack legal professional knowledge. Therefore, it's difficult for them to obtain and understand rights protection information, especially for migrant workers. Their legal consciousness was weak, without the awareness of preserving evidence in their work. They passively received legal aid after the conflict with employers intensified [26]. On the other hand, after returning to work, it is difficult for some people to obtain employment information. The pandemic inevitably impacted the traditional human resources service and intermediary platform, leading to severe principal-agent problems and time lag in employment information docking. However, the new Internet recruitment platform is more suitable for young people, which makes it more difficult for the elderly and low-skilled workers who are important customer groups of the traditional labor dispatch companies to find employment information [23]. According to the statistics of Zhilian recruitment platform, the number of job seekers aged 35 and above who submitted resumes on this platform increased by 14.9% year on year from February to September 2020. Meanwhile, middle-aged, and older job seekers faced a higher risk of long-term unemployment, and 62.9% of job seekers who resigned in March 2020 still submitted resumes in September [43].

6.4   Suggestions

To establish a long-term mechanism to protect the rights and interests of vulnerable groups, we need to continue to work from the following aspects to help vulnerable groups in the future.

In terms of the policies and regulations, it is necessary to improve policies and regulations on the protection of information service rights and interests. In China, the "Regulations on the construction of barrier-free environment" has been issued, which requires the governments at or above the county level and their relevant departments should create conditions to provide voice and text prompts and other information exchange services for the disabled when they release important government information and information related to the disabled. However, the sudden nature of the pandemic makes it difficult to fully implement the policy. It's necessary to develop new mechanisms for emergencies. For example, on February 18, 2020, the National Health Commission of the People's Republic of China (NHCPRC) issued a notice on the management of severe mental disorders during the pandemic. It explicitly requested that for the patients living in the quarantined area, it is necessary to provide door-to-door medicine delivery, network diagnosis and treatment and other services to ensure the patients' home treatment [31]. On March 13, 2020, NHCPRC issued another new notice to strengthen further the prevention and control of infection in medical institutions during the epidemic period so as to minimize cross-infection and protect medical staff and patients [32]. In October 2020, the China Disabled Persons' Federation issued the guidelines for social support services for the prevention of major infectious diseases for the disabled (for Trial Implementation), in which it is recommended to provide on-site rehabilitation training, remote diagnosis and treatment, supervision of medication, and substitution of medication for disabled people who need long-term rehabilitation treatment and are not able to move [7]. In response to emergencies, relevant government departments timely formulate and improve policies and regulations, improve the supervision mechanism, and strengthen the implementation of policy documents. Policy documents should be open and transparent in a timely manner. While actively using the mainstream media and the Internet, it is also necessary to make proper use of the traditional paper media to reduce the barriers brought by the digital divide and the cost of digital access to vulnerable groups, and fully protect the basic rights and interests of vulnerable groups.

In terms of information technologies and information systems, it is necessary to strengthen the research and development of barrier-free technology and products for vulnerable groups. With 5G, Internet of things, big data and other information technologies, it's necessary to develop more concise and friendly information tools for epidemic prevention, combined with an intelligent guide, voice assistance, intelligent live subtitles and other technologies to facilitate the access of information for the disabled. Nevertheless, for the vulnerable groups who do not have smart phones, cannot use smart phones, or cannot apply for health codes, we should provide alternative solutions to assist them in applying for relevant certificates to ensure that they can travel normally. In addition, it is necessary to strengthen the development of the information sharing platforms. During the outbreak of the epidemic, Wuhan once established the "epidemic prevention and control headquarters" for overall planning. Relying on the OA information platform, according to the information collected from community grid members and hospitals, patients were allocated to ensure the monitoring and timely intervention of mild patients, as well as the allocation of medical resources for patients with basic diseases. Resources were concentrated to treat severe patients, ensuring the efficient allocation of medical resources [64]. In the future, it is necessary to strengthen further the whole chain management of public health assistance at the national and even global level. In addition, from the perspective of epidemic prevention information tools, we should strengthen the interconnection and integration of different system tools, such as the cross-provincial health code, cross restricted sharing and mutual recognition of vaccination to reduce the travel barriers of vulnerable groups.

In terms of service content, it is necessary to establish and improve personalized information service content for different vulnerable groups. For example, establish mental health information service network for different groups. On April 7, 2020, novel coronavirus pneumonia, isolation workers and family members' psychological counseling and social work service program were issued by NHCPRC [30], which required extensive scientific education and mental health professional knowledge to help patients, segregated personnel, and their families to adjust their mental self. Communities, streets, and universities have also carried out public mental health information services (Jiangbei district society organization service center, [18, 28]. For another example, in the process of social and economic recovery, low-income and unemployed people are in urgent need of employment information. The Chinese government is taking preferential measures such as social security free relief and employment subsidies to help small, medium, and micro enterprises tide over the difficulties. At the same time, online employment services and skills training are also actively carried out. As of April 2020, the total number of online real-name registration trainings has been about 3.5 million [37]. Provincial and municipal government departments solve the employment problem by means of local placement and online recruitment. To better provide information navigation for vulnerable groups, public cultural service departments such as libraries can regularly collect these social public information service items and policies, sort out and archive them from multiple perspectives such as themes and users, to provide navigation and retrieval. All these measures could facilitate the information use of vulnerable groups. In the meantime, we should also strengthen health information literacy education and training for vulnerable groups and improve their awareness and ability to obtain health information.

In terms of operating mechanism, the ideal public information service system is often an open social system with the government as the core and the participation of enterprises and the third sector [53]. During the pandemic, government departments may be unable to provide perfect public information services due to lack of workforce, unclear responsibility subject, lack of emergency response plan, and untimely policy formulation and implementation. Therefore, encouraging public welfare organizations or non-profit organizations to carry out public information services actively can effectively make up for this deficiency. Encouraging the third party to provide public information can also ease the concerns of vulnerable groups to seek help from government departments. Government departments and public welfare organizations can jointly establish an emergency response mechanism to mobilize more social forces to help vulnerable groups. For example, some foundations such as Fujian Zhengrong Public Welfare Foundation, Zhejiang lakeside magic bean Foundation and Beijing Xiaogeng foundation for the disabled, actively participate in the prevention and control of the epidemic through fund-raising, voluntary service and rushing to the front line of medical care [57].

6.5   Conclusion

The vulnerable group is a relative concept, which will change with the social environment. It's necessary for countries all over the world to constantly assess which members of society are vulnerable groups to help the people at the highest risk fairly. Based on related vulnerable groups theory, we identified two types of vulnerable groups during the COVID-19 pandemic. By analyzing the difficulties of vulnerable groups in accessing the public information services during the pandemic, we put forward some suggestions. The limitation of this study was that we didn't collect data from vulnerable people. We only focused on one country and analyzed news reports and literature. Although the findings are enlightening to some extent, we need to conduct more in-depth research on the vulnerable groups caused by public emergencies and establish closer cooperation at the national and global levels to help more vulnerable groups obtain equal public information services.

References

1. Baig, Jalal. (2021, February 28). Cancer Patients Often Can't Get Full Care with Covid-19 Bogging down Medical Facilities. Retrieved from https://www.washingtonpost.com/health/cancer-patients-covid-effect/2021/02/26/cab3c26c-7608-11eb-9537-496158cc5fd9_story.html
2. Berg, J., Hilal, A., El, S., & Horne, R. (2021). World employment and social outlook: trends 2021. Retrieved from https://www.voced.edu.au/content/ngv:90668
3. Biu. (2021, November 30). Create AR Glasses Able to Read the World. Retrieved from https://baijiahao.baidu.com/s?id=1717847782961899530&wfr=spider&for=pc
4. Chang, J. (2020). Special Protection of the Human Rights of the Four Vulnerable Groups Under the Sudden Major Epidemic Situation. Journal of Human Rights, 2020(01),5–12. DOI: CNKI: SUN: RQYJ.0.2020-01-003

5. Chen, Q. (2020, February 24). The price of Wuhan community group purchase is high single category? Related parties respond to the six major problems. Retrieved from https://m.nbd.com.cn/articles/2020-02-24/1411103.html

6. Chen, X., M. (2021). Difficulties and Strategies in Obtaining Medical Treatment for Special Patients in Wuhan Under covid-19 epidemic. Medicine and Society (06),1–5. DOI: https://doi.org/10.13723/j.yxysh.2021.06.001

7. China disabled persons' federation. (2020, October 20). Notice on Printing and Distributing the Guidelines for Social Support Services for the Prevention of Major Infectious Diseases for the Disabled (for Trial Implementation). Retrieved July 22, 2021, from https://www.cdpf.org.cn/zwgk/ggtz1/c2b8757d2b304f8fba1419bead4af39e.htm

8. Chinese ministry of civil affairs. (2020, April 10). The Ministry of Civil Affairs Answered the Message of Netizens about "Resettlement of Stranded and Homeless People." Retrieved July 29, 2021, from http://www.gov.cn/hudong/2020-04/10/content_5500988.htm

9. Deng, Y. (2020, February 12). Wuhan First Hospital was requisitioned as an intensive care hospital for the new crown pneumonia, Protective material emergency. Retrieved from https://www.thepaper.cn/newsDetail_forward_5939155

10. Deutsche Welle. (2021, January 26). Coronavirus: Global GDP to Sink by $22 Trillion over COVID, Says IMF. Retrieved July 21, 2021, from https://www.dw.com/en/coronavirus-global-gdp-to-sink-by-22-trillion-over-covid-says-imf/a-56349323

11. Dong, B. (2020). Suspension treatment leads to disputes to protect workers. How should enterprises pay the bill. Enterprise Observer (03),22–23. DOI: CNKI:SUN:QYGC.0.2020-03-007

12. Fan, B. (2004). Empowerment of vulnerable groups and their model choices. Academic Research (12),73–78. DOI: CNKI:SUN:XSYJ.0.2004-12-00C

13. Fei, W. (2020, March 25). A survey of urban children's living and learning conditions during the epidemic. Retrieved July 6, 2021, from https://www.thepaper.cn/newsDetail_forward_6605896

14. Fung, H. F., Martinez, L., Alarid-Escudero, F., Salomon, J. A., Studdert, D. M., Andrews, J. R., ... & Ryckman, T. (2020). The Household Secondary Attack Rate of Severe Acute Respiratory Syndrome Coronavirus 2 (SARS-CoV-2): A Rapid Review. Clinical Infectious Diseases,73(Supplement_2), S138-S145. DOI: https://doi.org/10.1093/cid/ciaa1558

15. Huang, M. J., Qiu, Y.C., et al. (2020). Analysis on Prevention and Control Measures and Medical Resources of Obstetrical Hospital During the Epidemic Period of COVID-19 in Guangdong Province. The Journal of Practical Medicine(10),1277–1281

16. Hubei Provincial People's Government. (2020, February 17). Comments on the all-out effort to make sure the epidemic prevention and control. Retrieved from http://www.hubei.gov.cn/zwgk/hbyw/hbywqb/202002/t20200217_2039358.shtml

17. International Council of Nurses. (2021, January 13). INTERNATIONAL COUNCIL OF NURSES COVID-19 UPDATE. Retrieved from https://www.icn.ch/sites/default/files/inline-files/ICN%20COVID19%20update%20report%20FINAL.pdf

18. JiangBei district society organization service center. (2020, February 10). Online psychological support services are involved in advance, and Jiangbei's "management and love" pays attention to the mental health of home observation and isolation personnel. Retrieved from http://www.nbshzz.org.cn/cat/cat929/con_929_29025.html

19. Jing, Q. L., Liu, M. J., et al. (2020). Household secondary attack rate of COVID-19 and associated determinants in Guangzhou, China: a retrospective cohort study. The Lancet Infectious Diseases, 20(10), 1141–1150. DOI: https://doi.org/10.1016/S1473-3099(20)30471-0

20. Kang, L., Ma, S., Chen, M., Yang, J., Wang, Y., Li, R., .. & Liu, Z. (2020). Impact on mental health and perceptions of psychological care among medical and nursing staff in Wuhan during the 2019 novel coronavirus disease outbreak: A cross-sectional study. Brain, behavior, and immunity, 87, 11–17. DOI: https://doi.org/10.1016/j.bbi.2020.03.028

21. Lancet, T. (2020). Redefining vulnerability in the era of COVID-19. Lancet (London, England), 395(10230), 1089. DOI: https://doi.org/10.1016/S0140-6736(20)30757-1

22. Lei, K. (2020, February 17). Leaving a lifeline for critically ill patients with non-COVID 19. Retrieved from http://www.xinhuanet.com/comments/2020-02/17/c_1125584130.htm

23. Li, B., & Sun, Y.(2020). Analysis and Suggestions on Difficulties in Responding to the New Coronavirus Epidemic and Promoting Employment. China Employment(06),17–18. DOI: https://doi.org/10.16622/j.cnki.11-3709/d.2020.06.005

24. Li, J. (2020, March 6). The Ministry of Civil Affairs replied to the netizen's message about "stranded people, vagrants' placement". Retrieved from http://news.sina.com.cn/c/2020-03-06/doc-iimxyqvz8380058.shtml

25. Li, J., Luo, J., Li, Y., Guo, D., & Li, X. (2020). The influence of internet epidemic information causes cyberchondria among medical staff during COVID-2019 epidemic period. Psychologies (18),5–7+10. DOI: https://doi.org/10.19738/j.cnki.psy.2020.18.002

26. Li, W.(2020, May 30). Legal aid has successfully defended the rights of migrant workers by more than 510,000 people. Retrieved from https://www.creditchina.gov.cn/home/zhuantizhuan lan/nmgzt/nmgcase/202006/t20200601_197718.html

27. Lin, Y., Li, X., & Li, Y. (2018). Empirical Research on Employment and Overwork Situation of Couriers——Based on the Investigation of 1214 Couriers in Beijing. China Business and Market (08),79–88. DOI: https://doi.org/10.14089/j.cnki.cn11-3664/f.2018.08.009

28. Liu, C. (2020, February 2). Prevention and control of epidemic situation! National colleges and universities open psychological support hotline assembly. Retrieved from edu.china.com.cn/2020–02/02/content_75665451.htm

29. Moraliyage, H., Silva, D. de, Ranasinghe, W., Adikari, A., Alahakoon, D., Prasad, R., et al. (2021). Cancer in Lockdown: Impact of the COVID-19 Pandemic on Patients with Cancer. The Oncologist, 26(2), e342-e344. DOI: https://doi.org/10.1002/onco.13604

30. National Health Commission of the PRC. (2020, April 7). Notice on Printing and Distributing the Psychological Counseling and Social Work Service Scheme for Patients, Isolated Persons and Their Families in Covid-19. Retrieved from http://www.mca.gov.cn/article/xw/tzgg/202004/20200400026727.shtml

31. National Health Commission of the PRC. (2020, February 17). Notice on Strengthening the Treatment and Management of Patients with Severe Mental Disorders during the Covid-19 Epidemic. Retrieved July 22, 2021, from http://www.gov.cn/xinwen/2020-02/19/content_5480748.htm

32. National Health Commission of the PRC. (2020, March 13). Notice of the National Health Commission of the People's Republic of China on Further Strengthening the Prevention and Control of Infection in Medical Institutions during the Epidemic Period. Retrieved July 29, 2021, from http://www.gov.cn/zhengce/zhengceku/2020-03/13/content_5491044.htm

33. Nie, M. (2020, April 22). Outbreak in Hunan Women's Prison? Here comes the briefing!. Retrieved from www.chinapeace.gov.cn/chinapeace/c100054/2020-04/22/content_1234 2570.shtml

34. Nirappil, F., & Wan , W. (2021, January 6). Los Angeles Is Running out of Oxygen for Patients as Covid Hospitalizations Hit Record Highs Nationwide. Retrieved from https://www.washin gtonpost.com/health/2021/01/05/covid-hospitalizations-los-angeles-oxygen/

35. Office of the CPC Ministry of Industry and Information Technology of the PRC. (2020, March 1). All-out efforts during the epidemic prevention and control ensure the orderly and strong protection of medical supplies. Retrieved from http://www.qstheory.cn/dukan/qs/2020-03/01/c_1 125641807.htm
36. Pan, Y. (2016, March 30). Nearly 16% of China Population Suffer from Hearing Impairment. Retrieved from http://www.people.com.cn/n1/2016/0330/c32306-28238156.html
37. Party group of the ministry of human resources and social security of CPC. (2020, April 1). Go All out to Do a Good Job in Responding to the Epidemic and Stabilizing Employment. Retrieved from http://www.qstheory.cn/dukan/qs/2020-04/01/c_1125791178.htm
38. Polantz, K. (2020, August 3). Deaf Association Sues to Force White House to Use Sign Language Interpreters at Coronavirus Briefings. Retrieved from https://edition.cnn.com/2020/08/03/politics/sign-language-interpreters-coronavirus-briefings/index.html
39. Qi, W. (2020, March 6). Qi Weiei: Why are community workers busy, but Wuhan residents do not feel it?. Retrieved from https://www.guancha.cn/QiWeiWei/2020_03_06_540147_1.shtml
40. Qu, J. (2020, February 3). China Association of the Deaf and Hard of Hearing suggests that press releases have sign language interpretation or subtitles. Retrieved from http://ie.bjd.com.cn/5b165687a010550e5ddc0e6a/contentApp/5b16573ae4b02a9fe2d558f9/AP5e37de62e4b07 68ce199d0bc.html
41. Shi, Y., Wang, J., et al. (2020). Knowledge and attitudes of medical staff in Chinese psychiatric hospitals regarding COVID-19. Brain, Behavior, & Immunity-Health, 4, 100064. DOI: https://doi.org/10.1016/j.bbih.2020.100064
42. Stevenson, E., Barrios, L., et al. (2009). Pandemic influenza planning: addressing the needs of children. American Journal of Public Health, 99(S2), S255-S260. DOI: https://doi.org/10.2105/AJPH.2009.159970
43. Sun, T. (2021, January 25). The "dilemma" of re-employment of middle-aged and old-age job seekers: 60% of the job-seeking period exceeds half a year. Retrieved July 17, 2021, from http://news.bandao.cn/a/458361.html
44. Wang, M. (2021, May). Forced to go online? New group of seniors become active Internet users. Retrieved from http://epaper.ynet.com/html/2020-05/08/content_352853.htm?div=-1
45. Wang, R. (2020, March 3). If passengers stranded in Wuhan encounter temporary living difficulties, each person can receive 300-yuan assistance per day. Retrieved from http://www.hubei.gov.cn/zhuanti/2020/gzxxgzbd/qfqk/202003/t20200303_2171575.shtml
46. World health organization. (2021). WHO Coronavirus (COVID-19) Dashboard. Retrieved from https://covid19.who.int/ (accessed 18 July 2021)
47. Wu, D. & Zhang, C. (2020). Investigation of Health Information Literacy Among Patients Tested Positve with COVID-19 Infection and with Treatment. Library Journal (07),70–82. DOI: https://doi.org/10.13663/j.cnki.lj.2020.07.008
48. Wu, K. (2021). Research on the Protection of Rights and Interests of Express Employees. Co-Operative Economy & Science (01),186–187. DOI: https://doi.org/10.13665/j.cnki.hzjjykj.2021.01.077
49. Wu, M. (2020, February 28). The call for help burst, what is the difficulty of Wuhan rescue stranded people. Retrieved from https://www.yicai.com/news/100526016.html
50. Wu, W. (2020, November 19). What if you are dismissed and your salary is reduced during the epidemic? Beijing publishes typical cases and official explanations. Retrieved from https://www.bjnews.com.cn/detail/160575535815828.html
51. Wu, W., & Tian, C. (2020, April 8). More than 11,000 people stranded from Beijing in Wuhan will return to their hometown. Retrieved from http://www.xinhuanet.com/fortune/2020-04/08/c_1125829714.htm

52. Wu, Z., & McGoogan, J. M. (2020). Characteristics of and important lessons from the coronavirus disease 2019 (COVID-19) outbreak in China: summary of a report of 72 314 cases from the Chinese Center for Disease Control and Prevention. Jama, 323(13), 1239–1242. DOI: https://doi.org/10.1001/jama.2020.2648

53. Xia, Y. (2004). Social Choice of Public Information Service —— Analysis of the Relationship between Government and Third Sector Public Information Service. Journal of Library Science in China (03), DOI: CNKI:SUN:ZGTS.0.2004-03-002

54. Xiao, W., He, J., et al. (2020). Transmission Risks and response strategies for land transportation under the epidemic of coronavirus disease 2019. Injury Medicine (Electronic Edition) (02),1–8. DOI: CNKI:SUN:SHYD.0.2020-02-001

55. Xie, J. (2021). Suicidal ideation intervention counseling for children in epidemic situations. Mental Health Education in Primary and Secondary School (15),46–49

56. Xie, R., Xue, F. & Kan, Jian, L. (2011). Characteristics analysis of information need of clinicians during public health emergencies of infectious diseases. Chinese Journal of Health Education (06),451–453. DOI: https://doi.org/10.16168/j.cnki.issn.1002-9982.2011.06.022

57. Xing, C. (2020, April 5). Besides Donations and Materials, What Other Public Welfare Projects Do the Foundation Have to Fight against the Epidemic? Retrieved from https://baijiahao.baidu.com/s?id=1663069359523770798&wfr=spider&for=pc

58. Yan, X., Liu, X., et al. (2016). An outbreak of influenza in a psychiatric hospital. Chinese Journal of Infection Control (03),201–203. DOI: CNKI:SUN:GRKZ.0.2016-03-016

59. Yang, L. & Zhu, W. (2021). Analysis of prevention and control public opinions monitoring during COVID-19epidemic. Chinese Journal of Public Health Management (02),276–278. DOI: https://doi.org/10.19568/j.cnki.23-1318.2021.02.0037

60. Yao, H., Chen, J. H., & Xu, Y. F. (2020). Patients with mental health disorders in the COVID-19 epidemic.The Lancet Psychiatry, 7(4), e21. DOI: https://doi.org/10.1016/S2215-0366(20)30090-0

61. Ye, X. (2020, July 16). Face recognition payment, verification code, who can pave the "blind walkways" on the smartphone for the visually impaired Retrieved from http://ie.bjd.com.cn/5b165687a010550e5ddc0e6a/contentApp/5b16573ae4b02a9fe2d558f9/AP5f0fcbf4e4b086b26e0a885a.html?isshare=1

62. Yu, Y. & Yu, N. (2020). Practice and discussion on children health communication in public health emergencies. Chinese Journal of School Health (08),1124–1127. DOI: https://doi.org/10.16835/j.cnki.1000-9817.2020.08.002

63. Zarocostas, J. (2020), "How to fight an infodemic?", The Lancet, Vol.395 No.10225, pp.676. DOI: https://doi.org/10.1016/S0140-6736(20)30461-X

64. Zhang, D., Huang, J., Luo, C. (2021). From Hospital Runs" to Health Care for All": The Impact of the Public health Policy Change. China Journal of Economics (02),182–206. DOI: https://doi.org/10.16513/j.cnki.cje.20210602.003

65. Zhang, W. (2020, February 15). How can we help the handicapped during the epidemic? Retrieved from. https://rmh.pdnews.cn/Pc/ArtInfoApi/article?id=11531548

66. Zhang, Z. & Ren, W. (2020, March 2). Stranded outlanders in Wuhan. Retrieved July 21, 2021, from https://www.thepaper.cn/newsDetail_forward_6238991

67. Zhao, Y. (2021). Analysis of the dilemma of information disadvantaged groups under digital governance. Co-Operative Economy & Science (09),184–187. DOI: https://doi.org/10.13665/j.cnki.hzjjykj.2021.09.075

68. Zheng, G. (2020, March 30). Investigation on the Group of Couriers and Takeaways in Wuhan (I): Work and Life before and after the Epidemic. Retrieved from https://www.thepaper.cn/newsDetail_forward_6733192_1

69. Zhu, J. H., Li, W., et al. (2021). The Attitude towards Preventive Measures and Knowledge of COVID-19 Inpatients with Severe Mental Illness in Economically Underdeveloped Areas of China. Psychiatric Quarterly, 92(2), 683–691.DOI: https://doi.org/10.1007/s11126-020-098 35-1

70. Zhu, M. (2020, July 15). Do not let the "scan code difficulties" become the death of the digital age for the elders. Retrieved from http://yuqing.people.com.cn/n1/2020/0715/c209043-317 84748.html

71. Zhu, W., Zhang, P., Li Peng, F. & Wang, Z. (2020). Firm Crisis, Government Support and Policy Efficiency under the Epidemic Shock: Evidence from Two Waves of Questionnaire on SMEs. Management World (04),13–26. DOI: https://doi.org/10.19744/j.cnki.11-1235/f.2020.0049

72. Zhu, Y. (2020, March 6). Face to face with the epidemic prevention and control: Wuhan residents' livelihood security issues are of concern. Retrieved from http://sd.people.com.cn/GB/n2/2020/0309/c373025-33860721.html

Kun Huang, Ph.D., is a Professor at the Department of Information Management, School of Government, Beijing Normal University, Beijing, China. She broadly examines the difficulties of information seeking in the context of COVID-19 and discusses the challenges the vulnerable groups face in information seeking.

She has secured research grants and contracts, including from the National Natural Science Foundation of China and the Humanities and Social Sciences Foundation of the Ministry of Education. Her work is published in various journals, and she has presented at several conferences related to information behavior, affective information processing, and digital library.

Dr. Huang received her Ph.D. from Peking University in Management. She received her M.S. in Management and B.S. in Science from Beijing Normal University. She is a member of Information Behavior Research Committee of China Science and Technology Information Society. She is also a member of Statistics and Evaluation committee of Chinese Library Society.

Lei Li, Ph.D., is an Assistant Professor of the Department of Information Management at the School of Government, Beijing Normal University, Beijing, China. She obtained her PhD in management science and engineering from Nanjing University of Science and Technology in China. Her research interests involve user information behavior, information quality judgement, text mining and social media. She chairs a National Social Science Foundation of China. Her research has been published in journals including Information Processing & Management, Journal of the Association for Information Science and Technology, Aslib Journal of Information Management, Online Information Review, Library Hi Tech, The Electronic Library and in the proceedings of some international conferences.

Shichao Luo is a graduate student of Library and Information Science student in the School of Government, Beijing Normal University. He has examined and integrated the news and reports about challenges of accessing the public information service the COVID-19 brought to the Vulnerable Groups. Luo received his bachelor's degree from Nanchang University in Archival Science. He has two years of work experience in the department of local government, including the Market Supervision Bureau and the Chorography Office in Ganzhou, Jangxi Province. He participated in the poverty alleviation and the epidemic prevention.

Xiaoyu Wang is a graduate student of Library and Information Science student in the School of Government, Beijing Normal University. Her research interest is information behavior. She acquires and collects the situation of public information dissemination and acquisition under crisis and the unfavorable factors for vulnerable groups, including news reports, academic papers, and monographs.

She received her bachelor's degree from Liaoning University in Archival Science. She got excellent grades in undergraduate study and was recommended to study for a master's degree without examination. She has participated in many social welfare service activities and anti-epidemic work.

Responding to COVID-19: Privacy Implications of the Rapid Adoption of ICTs

Thora Knight, Xiaojun Yuan, DeeDee Bennett Gayle, and Salimah LaForce

Abstract

COVID-19 increased reliance on information communication technologies (ICTs) as public and private organizations altered standard business operations to adhere to public health guidance. Across most sectors, technology deployment was swift, which left organizations with little opportunity to assess corresponding impacts. This chapter highlights various technologies implemented during the pandemic within four key sectors, government, education, healthcare, and employment, and the purpose these technologies serve. Social implications of the widespread use of these technologies are discussed, emphasizing privacy, trust, ethics, and potential effects on socially vulnerable populations. Through the Company Information Privacy Orientation (CIPO) privacy framework, this chapter also presents factors that public and private organizations should consider in emergency technology deployment. The chapter closes with research considerations to further understand the role of ethics, privacy, and trust in using ICTs, to facilitate core functions of life, which will continue after the pandemic ebbs away.

Keywords

Privacy • Ethics • Trust • Information and Communication Technology • COVID-19

T. Knight (✉) · X. Yuan · D. Bennett Gayle
University at Albany, State University of New York, Albany, NY, USA
e-mail: tknight@albany.edu

S. LaForce
Georgia Institute of Technology, Atlanta, GA, USA

© The Author(s), under exclusive license to Springer Nature Switzerland AG 2023   107
X. Yuan et al. (eds.), *Social Vulnerability to COVID-19*, Synthesis Lectures
on Information Concepts, Retrieval, and Services,
https://doi.org/10.1007/978-3-031-06897-3_7

7.1   Introduction

The primary public health response to the COVID-19 pandemic was to 'flatten the curve.' Flattening the curve described cooperative societal efforts to reduce the number of people needing hospitalization and medical care, thereby limiting the spread of the virus. At the start of the pandemic, most states in America and countries worldwide controlled citizens' movements by requesting that non-essential employees work from home and for K-12 schools and universities to transition temporarily to remote online learning. Further, government, emergency services personnel, employers, and healthcare officials implemented processes to reduce potential exposure to the virus. Across most sectors, technology and broadband wireless were essential to continue societal functions. The responses to the COVID-19 pandemic exposed (among other things) people's increasing reliance on technology.

As explained in previous chapters, social vulnerability is a term to group populations based on their situation which may cause difficulties prior to, during and after a disaster. The difficulties are primarily due to these populations having less access to resources, finances to purchase resources, or power to demand access [16, 62]. One such resource is broadband internet access, and the technologies that rely on internet access. Though disaster studies have examined the impact of the lack of access to these technologies, public health researchers have, as well.

According to Benda et al. [5], as a social determinant of health, broadband internet access was important during COVID-19, and lack of broadband internet access influenced each of the six social determinant of health domains identified by the American Medical Association [6], as well as an additional domain, access to credible information (see Fig. 7.1).

While the populations included in the broad term 'socially vulnerable' are varied by region or country, the types of accessibility issues differ across sector, as well. For example, during the COVID-19 pandemic, older adults were in triple jeopardy in comparison to younger people [86, 87]. Not only were they at higher risk of contracting the virus, they

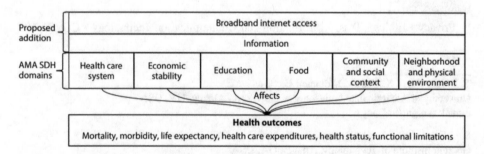

Fig. 7.1 Proposed Extended Model of Social Determinants of Health (SDH) (adopted from Benda et al. [5], source from American Medical Association [6]

were also at higher risk of mortality due to underlying conditions. Additionally, older adults were more likely to have difficulties moving online for employment, healthcare, government services, or access to specialized organizations. [21] mentioned two reasons that older adults were deprived of technological advances, including "our inherent bias of assuming the aging population is passive and lacks the ability to learn," and "this is a population who does not advocate for itself." Eghtesadi [21] further requested that "we must update our practice of medicine and integrate assessment of technology use as part of the preventative healthcare we offer to vulnerable populations." (p. 949).

In another example, Dubois et al. [17] examined the technology adoption of racial and ethnic minority students in the United States during the COVID-19 pandemic and found that "technology literacy is a necessity for civic and economic activity," and "teachers validate digital competency differently depending on students' race and class" (p. 14). They also mentioned that "Blacks in urban public schools have often been deprived of the resources and opportunities to engender technological ambition." (p. 15).

This chapter highlights various sector adoption of technology and corresponding (social vulnerability) concerns that are important to consider when understanding the complexity of the COVID-19 disaster response. This chapter consists of two main sections. In the first part, the critical reliance of technological deployments is discussed as a function of sector involvement, segmented by government, education, healthcare, and employment (highlighting the social vulnerability concerns of various populations), followed by a discussion on the privacy and security implications of the rapid adoption of technology and the potential impact to socially vulnerable populations in the second part.

7.2 Technological Deployments

The global COVID-19 pandemic caused significant changes to people's everyday lives. States issued executive orders to close public schools, and universities required non-essential personnel to work from home. As a result, technology played a more significant role in every aspect of people's lives. This section explores the technological deployment in government, education, employment, and healthcare in response to COVID-19.

7.2.1 Government

The response to COVID-19 led to a shift in government delivery of services, including emergency response initiatives. Efforts to maintain 'social distancing, 'flatten the curve,' and reduce density ushered in new or expanded governmental uses of technology. To keep public services running and to manage the pandemic, governments implemented various technology-supported initiatives. Several technologies emerged as critical to government management. Unmanned Aerial Vehicles (UAVs) stood out for their multipurpose uses. Because UAVs are operated remotely, without direct human touch, and function autonomously, governments found them an appropriate technological deployment. In Japan, UAVs were used in airports to identify potentially infected citizens [74]. In Italy and Morocco, they monitored the streets and issued warnings to people found breaking lockdown rules in addition to identifying those with elevated temperatures [49, 54]. In the U.S., at least on state, Connecticut, considered using UAVs to monitor beaches, train stations, recreation areas, and shopping centers [55]. However, law enforcement abandoned the pilot program citing public concerns surrounding privacy and the effectiveness of technology in achieving its deployed purpose [9, 55]. In North Carolina, the state transportation department used UAVs to deliver protective equipment and food in a public-private partnership [33]. A less common use of UAVs surfaced in Ghana, where they were used to deliver millions of doses of COVID-19 vaccines with plans to deliver to remote and roadless areas by the end of 2021 [18, 31]. Like UAVs, other robots were used to promote contactless interaction and governments used them to broadcast public service announcements (PSAs), enforce social distancing rules, and disinfect facilities [49, 71]. Previous research findings in US and Canada suggest that communities that have tensions with law enforcement may feel further marginalization regarding the use of UAVs by these agencies [60]. Furthermore, the perception of privacy among marginalized populations may differ by country [25].

Though country-by-country adoption of contact tracing technology varied, it nonetheless became part of the public sector's arsenal in mitigating the spread of COVID-19. Contact tracing is the process of contacting people who were near someone who was diagnosed with the COVID-19 virus. Traditionally carried out in-person, digital contact tracing supplemented in-person measures for countries where participation was voluntary and served as the primary mode of tracing in countries where mandated. Over 44 countries adopted contact tracing technology via mobile applications utilizing Bluetooth, GPS, and location technologies [51, 64]. At least one country, Singapore, provided wearable contact tracing devices in lieu of installing the mobile app [88]. Some contact tracing initiatives in the U.S. involved public–private collaboration. For example, instead of developing a new contact tracing infrastructure, California adopted the joint Exposure Notification framework created by Google and Apple [12]. Due to potential inequities regarding the use of digital contact tracing technologies the World Health Organization published guidance for its member countries and noted that without certain levels of trust in government the public may not voluntarily adopt such applications [85]. While levels of trust in government tend to be lower among marginalized populations, the WHO guidance also cautioned that in-country "inequities could be exacerbated through the use of these [digital contact tracing] technologies," primarily benefiting those who have high levels of digital literacy [85, p. 2].

Other government initiatives explored technologies for contactless transactions, including the use of facial recognition systems. Examples included a national security agency program that tested and evaluated touchless identification to prevent employees from having to touch documents from people seeking services [45]. Others included technologies for authentication and surveillance. For instance, in India, iris scans were expected to replace biometric fingerprinting for authentication at specific COVID vaccination sites (The Wire [75]; Qureshi [56]). Interest in facial recognition technologies was high despite some law enforcement departments announcing a ban or potential ban on the expansion of facial recognition programs—for instance, in Minnesota [39] and London [40]. Two reasons for these bans are: (1) the disparate impact on marginalized populations and (2) privacy concerns regarding ethics and consent [42, 57, 63].

Further, a Government Accountability [78] report published in 2021 found that 75% of the 24 U.S. agencies surveyed used facial recognition technologies in 2020, and 40% planned to expand their programs through 2023. Their facial recognition uses included unlocking agency-issued smartphones, identifying persons of interest and victims of crime, monitoring, and physical security [78]. The most noteworthy COVID-19 finding was the government's interest in research on facial recognition capability to identify people wearing masks [78]. Given the increased use of facial recognition, there has been a call for regulation, at least in the US [42].

7.2.2 Education

Continuation of formal learning during the pandemic was essential to ensure students met their educational milestones. Various technologies have been experimented to foster the transition to remote learning. Surma and Kirschner [72] asserted that certain fundamental learning and instruction principles must be employed regardless of the type of distance learning tools used because they are beyond the learning environment. In response to COVID-19, cloud-based Learning Management Systems (e.g., Google Classroom, Canvas) and digital learning tools (e.g., Edgenuity, Class DoJo, Remind) experienced an increase in use. Techniques such as gamification, sometimes referred to as edutainment, showed promise in not only engaging young students [46] but also can be used as a learning assessment tool. Despite concerns about screen time, some asserted that end-of-day debriefing dialogues are critical to reinforcing learning that occurred during the day and engaging students in a dynamic online discussion environment [10]. Video conferencing platforms to offer live instructions, and presumably, debriefing dialogues became commonplace during the COVID-19 pandemic. Though one study identified that the biggest obstacle to remote learning for post-secondary students was motivation, students from low-income households had more obstacles to cross [68]. The findings from a study on children in Nigeria were similar, students from low-income communities had more barriers to remote instruction at home [11].

Online learning could be synchronous or asynchronous. Given the diversity of students and educators, not to mention their learning and teaching styles, educators needed to think about how to optimally combine synchronous and asynchronous instruction. In higher education, researchers found that medical students and other hands-on professions compared to the social sciences/liberal arts would require different techniques to engage the students and enable learning of complex technical maneuvers and concepts [73, 77]. One study found that Google Classroom was an effective tool for college-level English teaching and learning [73]. The Neurosurgical Atlas, which includes a virtual reality platform and 3D modeling of anatomy, has experienced a more than 20% increase in traffic since COVID-19, and students have recognized virtual reality as superior to 2D textbook illustrations [77]. The proliferation of various educational tools would undoubtedly bring changes to how people receive formal learning. For example, a study in Israel identified potential barriers for students with learning disabilities and remote instruction [22].

7.2.3 Healthcare

Virtual health care expenses increased from about $3bn Pre-COVID-19 to about $250bn Post-COVID-19 as people relied more on telemedicine and telehealth for treatment and to communicate with their health care providers [7, 14, 20]. Rockwell and Gilroy [59] reported that telemedicine helped hospitals treat patients remotely through telehealth systems, videoconferencing, and mobile applications. Video consultations to treat patients with various levels of symptoms have become increasingly acceptable due to the pandemic [30]. In China, for instance, people were able to communicate with doctors available online in real-time on websites, such as QuickDoctorOnline, to schedule doctor appointments and obtain medical advice for their health concerns [89]. After hospitals and clinics reached patient capacity, the Chinese national health insurance agency provided financial support to enable doctors to implement virtual care consultations [81].

Emerging technologies like robots, the Internet of things (IoT), artificial intelligence, data science, deep learning, and blockchain technology are being adopted for patient care [76]. IoT was used for live tracking and real-time updates of COVID-19 cases in the United States, Singapore, the United Kingdom., and China [76]. According to Global-Data Healthcare [27], "Wearables [were] one of the fastest-growing sectors in the global technology landscape, and using wearable devices to track vital COVID-19 symptoms to identify patterns predicting onset is an attractive prospect" (para. 4). Northwestern University and Shirley Ryan AbilityLab in Chicago collaboratively developed a wearable device to detect early signs and symptoms of COVID-19 [47]. Meanwhile, research from Rockefeller Neuroscience Institute in the U.S. showed that the Oura ring, a wearable sleep and activity tracker, can be used to predict the early signs of COVID-19 [27]. Scripps Research Translational Institute used wearable devices such as Fitbit, Apple Watch, and Garmin to track potential cases of COVID-19 in the Digital Engagement & Tracking for Early Control & Treatment (DETECT) study. The results demonstrated that "Fitbits can predict COVID-19 in 78% of the 14 patients studied" [27, para. 2]. While this subset of IoT was useful for collection of data for analysis, blockchain technology was also employed to facilitate the delivery of medications to pharmacies, as well as to patients in their homes [27, 76].

Research continues regarding the use of robots for treating COVID-19 patients. The main types of robots used in health care included surgical robots, exoskeletons, care robots, and hospital robots [79]. It is worth mentioning that care robots provided"care and support to elderly and disabled patients" and were in great demand in countries such as Japan, "which [was] facing a predicted shortfall in the number of available caregivers" [79, para. 8]. Hospital robots can "deliver medications, laboratory specimens, and other sensitive material within a hospital environment" [79, para. 12]. Data science technologies, including databases, machine learning, and artificial intelligence, were being used to identify patterns as researchers work diligently to develop a vaccine for COVID-19 [58]. However, researchers may have found their efforts hinderedd by the limitations of these technologies. For example, when creating databases of literature and medical documents in different languages for current and future coronavirus research, researchers had to (and will have to) deal with the comprehensiveness and accuracy of the contents of these information systems. These research challenges are further complicated by existing questions about the trustworthiness, biases, and inaccuracies of algorithms that enable the deployment of artificial intelligence systems. Hospitals took advantage of the benefits of the large-scale computing capabilities that emerging technologies offer. Health information systems have seen dramatic changes since 1960, and hospitals deployed distributed computing systems and robust networks in their health information system setting [28]. In March 2020, Health Catalyst, Inc., a healthcare data analytics technology and services provider, offered three initial solutions to health system clients, including patient tracking, public health surveillance, and staff augmentation support in response to COVID-19 [32].

7.2.4 Employment

In April 2020, Ernst & Young (E&Y) and the International Association of Privacy Professionals (IAPP) surveyed 933 privacy professionals to understand how organizations responded to the COVID-19 pandemic [35]. Respondents were diverse, representing organizations located in over 80 countries, with about 50% of respondents outside the U.S.. People working in government, education, and health represented the top five job sectors [35]. The survey [35] revealed that 90% of the respondent's organizations required all or almost all their employees to work from home. In adopting the work-from-home model, 45% of these organizations adopted new technologies or worked with new technology vendors. Technologies adopted in the employment segment were consistent with emerging trends and included surveillance, health monitoring, communication, and artificial intelligence.

While employer use of tracking software that measured employee productivity increased before the pandemic [67], the shift to remote working accelerated demand. Employee monitoring technologies included popular software such as Slack which allowed supervisors to access employees' private messages [30]. They also included less visible brands such as Teamviewer, ActivTrak, Time Doctor, Teramind, and Hubstaff, which enabled employers to track the amount of time employees spent on work versus non-work activities. Some employee monitoring technologies included software that mirrored the employee's laptop from home and projected the image onto a desktop at the office, allowing a manager to have real-time access to the employees' activities [48]. Other technologies kept track of employee mouse movements, keyboard strokes, and the web pages they visited [2]. These type of monitoring systems may have negatively impacted women workers. During the pandemic, the burden for women increased, as many were struggling to balance work and family [1, 37]. The increased burden varied across countries, regions, and cultures as the social construction of gendered activities differed [1].

In addition to contact tracing apps, essential services employers Amazon and Walmart used infrared cameras to check employee's temperature and infection rate [15]. More advanced adoptions integrated artificial intelligence to monitor employee movements and behaviors. A notable example is a social distancing detection tool developed by Landing A.I. [41]. Employers could integrate the software into their security camera systems. It would then detect and notify employees who were not maintaining social distancing rules [41]. Other AI-supported technologies targeting health and manufacturing industries work with radio-frequency identification (RFID) embedded in lanyards, tracking badges, and mobile phone applications. Some of these systems were configured to collect minute-by-minute data and generate proximity reports of employee distance from each other [8].

Videoconferencing through Zoom became a staple pandemic technology for face-to-face meetings. However, the exponential growth Zoom experiences during the pandemic raised privacy and security issues about how the company managed users' electronic data [82]. Against the backdrop of public outcry and regulatory oversight, Zoom deployed various security features to remedy system vulnerabilities [80]. Amid calls for more robust security and privacy measures, Microsoft (Teams) partnered with Verint to allow companies to integrate compliance tools securely [84]. The compliance software allowed companies to capture, retain, analyze, and retrieve all communications from Microsoft Teams, including voice calling, chat, online meetings, and screen sharing logs to ensure business continuity and regulatory compliance [84]. This partnership demonstrated the data gathering capabilities of technologies, suggesting that different interests may be at play. In this example, data aggregation for business compliance and the privacy and security needs of people whose activities were recorded.

7.3 Implications of Rapid Technology Adoption

The review of the technological deployments in response to the pandemic sheds light on the value of information and emerging technologies to individuals, organizations, and society. It also emphasizes the versatility of various technologies adapted across sectors, as summarized in Table 7.1. Common functionality of technologies across sectors is the ability to collect, analyze, process and use large amounts of data about people in their various roles as citizens, students, patients, employees, and in fact, in every facet of our lives. Before the pandemic, privacy and security were known challenges inherent in using emerging information communication technologies. However, the rate and the scale of deployments in response to COVID-19 add new layers to the unintended consequences of the widespread use of digital technologies. This section discusses privacy and security concerns associated with technological adoption and the importance of accounting for these concerns in future disaster, emergency response planning.

Table 7.1 Summary of technology deployments and uses

Technology	Use
Artificial intelligence	Social distancing detection and management; COVID-19 research
Mobile applications and devices	Virus tracing and notification
Drones	Public hygiene, delivery of food and medicine; communication with the public, virus detection
Facial recognition	Identification and authentication of individuals
Gamification	Learning assessment
Internet-of-Things	Live-tracking and real-time updates on COVID-19, virus detection; contact tracing
Learning management systems	Online learning delivery
Location tracking (*GPS, Bluetooth, RFID*)	Social distancing detection and enforcement, employee tracking
Robots	Disinfecting public spaces, care, and support for the elderly and disabled
Employee monitoring	Monitor keystrokes view messages, access and view device in real-time
Telehealth/Telemedicine	Health visits—treatment and diagnosis
Videoconferencing	Live instruction, debriefing dialogues, meetings, health visits

7.3.1 Privacy, Ethics, Trust Issues in Information Technologies

Concerns about privacy and security in digital technologies are well documented. A 2000 Pew Research survey showed that Americans were concerned about privacy on the Internet. Among the top two concerns were the disclosure of personal information to people they do not know and that computer hackers could access their personal information [23]. Over the years, Pew Research has continued to survey the public on privacy and security in relation to big tech, government, business, and the pandemic more recently [53]. Survey findings on digital privacy issues suggest that people have concerns about companies using and collecting information about them. They are also concerned about the government collecting their information [4] and the effectiveness of location tracking technologies in mobile phones to limit the spread of the virus [3]. The latter survey also found that perceptions were mixed on whether it was acceptable for the government to track people who tested positive for the virus but were against using mobile tracking to ensure compliance with social distancing rules [3].

Privacy theories have established that data collection and data use, sharing, improper access, inaccurate data, control, and awareness are fundamental dimensions of privacy concerns in online information systems [44, 66]. The technologies discussed are all capable of collecting or processing data. While some data collection, such as dispensing disinfectants, is benign, surveillance in recreational areas and the workplace and mobile devices can be problematic. The question of trust arises when people's activities and behaviors are surveilled. Researchers have concluded that concepts of trust and ethics are interwoven with the concept of privacy with computer-mediated technologies [13, 38]. The authors made this assessment when studying the use of surveillance technologies in the workplace. Their study and others have found that constant use of intrusive tracking technologies can break down trust between employees and employers [13, 69]. Active monitoring using technologies that continuously collect, store, analyze, and report employee behavior, and performance has been found to reduce employee's perception of an organization [19, 34]. Lawsuits filed against Zoom emphasized security and data sharing concerns at the intersection of ethics and trust surrounding information sharing. In 2021, Zoom agreed to settle claims that alleged, among other concerns, that the company (1) shared personal information with social media websites for profiling and marketing purposes and (2) failed to implement proper security measures. [80, 83]. Therefore, privacy and security issues can occur both because of how technology is used and how data is managed.

7.3.2   Post-COVID Considerations

The COVID-19 pandemic was not a widely anticipated disaster leaving organizations across sectors unprepared to maintain normal functions. A key takeaway from the pandemic is that technology and corresponding privacy and security implications should be incorporated in business continuity and disaster response plan. The E&Y and IAPP survey revealed that of the privacy professionals surveyed, 60% indicated their companies bypassed privacy and security review when they adopted new work-from-home technologies [35]. Thus, even companies with personnel dedicated to implementing privacy compliance did not adhere to standard organizational practices and policies. Highly regulated industries such as health, finance, and education are likely to have privacy compliance programs. Such programs should incorporate emergency plans for deploying technologies during emergencies, including measures for evaluating the types of technologies required for different functions, the potential stakeholders who would be affected, and the types of personal data, if any, that would be implicated.

Stakeholder analysis is essential because research has shown that disasters tend to worsen pre-existing digital inequalities [36, 43]. In some but not all contexts, socially vulnerable populations' privacy and security needs may differ from the general population. For instance, older adults are less likely to have the requisite skills and knowledge to identify security threats and protect their information [24]. They are also more likely to need medical care. Given that privacy has been identified as a factor in older adults' adoption of technology [24, 52], addressing this concern would enhance the benefits that telehealth and telemedicine services can offer to this population. Similarly, privacy issues may arise when students are required to turn their cameras on during synchronous learning [61]. This simple requirement "can be especially stressful for disadvantaged students, students of color, those with disabilities; undocumented students; students in temporary living situations or those from low-income families, living in crowded homes or apartments" [70, para. 5]. Thus privacy rest not just in the collection of their information but also in the disclosure of their personal environment if classroom policies require them to reveal their surrounding in a synchronous learning platform. Given individuals' role vary according to the context, citizen, employee, patient, student, customer or vulnerability, the privacy and security analysis is complex, and the strategy employed would be one that is based on a continuum.

The Company Information Privacy Orientation (CIPO) provides an adaptable framework for organizations planning to deploy sophisticated information systems [29]. Although the authors created the framework with companies in mind, it is amendable to public sector technology deployments. They identified three dimensions of privacy challenges, information management strategy, ethics, and law. Ethical obligations refer to the organization's definition of its obligations to its customers [employees or citizens]. Information management strategy is the organization's purpose for gathering and using personal information. The four strategies identified in the literature are transaction-focused, data-focused, inference-focused, and advice-focused [29]. The corresponding purposes for gathering and using information include completing an immediate transaction, increasing profitability, improving organizational understanding about individuals, and providing individualized services [29]. Legal risk assessment is the organization's view of the impact that privacy laws have on its ability to carry out its [functions] [29]. The legal assessment refers to whether organizations view privacy laws as a constraint or an opportunity. However, in a post-COVID-19 privacy and technology analysis, the legal dimension would instead focus on integrating existing laws or industry self-regulatory frameworks into their emergency technology response plans.

The overarching theories of the framework are control theory and social justice theory. Control theory explains the extent organizations allow [individuals] to exercise the way their information is collected, used, and reused. Social justice theory explains the level of transparency organizations offer [29]. The ethical dimension based on three theories, stockholder, stakeholder, and social contract theory, defines the organization's philosophy on whose interests are to be served. Under stockholder theory, organizations are obligated to act in the best interests of stockholders, and the organization's interests are paramount [65]. On the other hand, stakeholder theory recognizes that multiple parties' interests are at stake and take a balanced approach to information privacy [65]. The final theory that is important to the ethical guidance to organizations' management of information is the social contract theory that posits that the [the individual whose data is being gathered] interests are paramount and that organizational actions and practices should not outweigh the benefits [50, 65]. This framework demonstrates that assessing the privacy implications of deploying technology even in regular times is complex. It requires analyzing ways to approach information management, legal factors that may guide data practices, and considering the parties whose information is at risk.

7.4   Conclusion

The individual and collaborative efforts of social sectors to flatten the curve and mitigate the spread of COVID-19 has undoubtedly increased individual and organizational reliance on technology. There is an ongoing question about whether this adjustment is a trigger for long-term structural change. The response to COVID-19 has exposed the integration and interaction among the human and physical environments. Technological developments have occurred in education, employment, government, and healthcare. The rapid and near-ubiquitous adoption of technology across these sectors has short- and long-term impacts on privacy, ethics, and trust, in information technology systems. Without adequate intervention, it will further pre-existing inequalities. While some of the impacts are novel to COVID-19, many are well documented throughout disaster research as occurring across the spectrum of all-hazards.

Future research should continue to explore the role of privacy, trust, and ethics in information technologies. In order to understand the technological impact of the pandemic, perception studies are needed to gain insights into how individuals view the collective impact of technology on various aspects of their lives. Studies should focus on the isolated sectors, such as how healthcare practitioners implement privacy policies or technical measures in telehealth, and should take a holistic approach to inform information systems. Studies should also capture viewpoints in dyadic relationships such as the practitioner and the patient. Similarly, studies that explore the impact of technology on employment should seek the viewpoint of both the employer and employee. Drawing upon the theories that underpin, the CIPO framework approaches should be employed in research, business, and government environments. This will ensure that the voices of populations affected by technology disparities are heard and included in studies at the intersection of technology adoption and COVID-19. Previous research studies should not be relegated to old books and history. Instead, the lessons learned regarding vulnerable populations and inequities should be incorporated into the future response to any hazard.

Acknowledgements Working Group [Thora Knight and DeeDee Bennett, Co-leads], "Technological Innovations during COVID-19". COVID-19 Working Group for Public Health and Social Sciences Research. Supported by the National Science Foundation-funded Social Science Extreme Events Research (SSEER) network and the CONVERGE facility at the Natural Hazards Center at the University of Colorado Boulder.

References

1. Aldossari, M., & Chaudhry, S. (2021). Women and burnout in the context of a pandemic. Gender, Work & Organization, 28(2), 826–834.
2. Allyn, B. (2020, May 13). *Your boss is watching you: Work-from-home boom leads to more surveillance.* NPR.Org. https://www.npr.org/2020/05/13/854014403/your-boss-is-watching-you-work-from-home-boom-leads-to-more-surveillance.
3. Anderson, M., & Auxier, B. (2021). Most Americans don't think cellphone tracking will help limit COVID-19, are divided on whether it's acceptable. *Pew Research Center.* https://www.pewresearch.org/fact-tank/2020/04/16/most-americans-dont-think-cellphone-tracking-will-help-limit-covid-19-are-divided-on-whether-its-acceptable/.
4. Auxier, B. (2020, May 4). How Americans see digital privacy issues amid the COVID-19 outbreak. *Pew Research Center.* https://www.pewresearch.org/fact-tank/2020/05/04/how-americans-see-digital-privacy-issues-amid-the-covid-19-outbreak/.
5. Benda, N. C., Veinot, T. C., Sieck, C. J., & Ancker, J. S. (2020). Broadband internet access is a social determinant of health!.
6. Bennett, N., Brown, M., Green, T., Hall, L., & Winkler, A. (2018). Addressing social determinants of health (SDOH): Beyond the clinic walls. Retrieved from American Medical Association: https://edhub.ama-assn.org/stepsforward/module/2702762.

7. Bestsennyy, O., Gilbert, G., Harris, A., & Rost, J. (2020). Telehealth: A post-COVID-19 reality? McKinsey. https://www.mckinsey.com/industries/healthcare-systems-and-services/our-insights/telehealth-a-quarter-trillion-dollar-post-covid-19-reality.

8. Bittle, J. (2020, May 5). *Companies Want to Use RFID Technology to Make Sure Employees Are Washing Their Hands*. Slate Magazine. https://slate.com/technology/2020/05/rfid-technology-handwashing-social-distancing-coronavirus-workplace.html.

9. Blair, R. (2020). Connecticut town scraps plan to use temperature-tracking drones in fight against coronavirus. Courant.Com. https://www.courant.com/coronavirus/hc-news-coronavirus-connecticut-drone-20200423-tiakrmg3erez7fpkpxk6ixkacy-story.html.

10. Bradley, C. S., Johnson, B. K., & Dreifuerst, K. T. (2020). Debriefing: a place for enthusiastic teaching and learning at a distance. Clinical Simulation in Nursing, https://doi.org/10.1016/j.ecns.2020.04.001.

11. Briggs, D. C. (2020). COVID-19: The Effect of Lockdown on Children's Remote Learning Experience–Parents' Perspective. *Journal of Education, Society and Behavioural Science*, 42–52.

12. California Department of Public Health. (2021, June 30). *Contact Tracing Q&A*. https://www.cdph.ca.gov/Programs/CID/DCDC/Pages/COVID-19/Contact-Tracing-QA.aspx#canotify.

13. Carew, P. J., Stapleton, L., & Byrne, G. J. (2008). Implications of an ethic of privacy for human-centered systems engineering. A.I. & Society, 22(3), 385–403. https://doi.org/10.1007/s00146-007-0149-7.

14. Centers for Medicare & Medicaid Services. (2020). State Medicaid & CHIP Telehealth Toolkit-Policy Considerations for States Expanding Use of Telehealth. https://www.medicaid.gov/medicaid/benefits/downloads/medicaid-chip-telehealth-toolkit.pdf.

15. Chyi, N. (2020, May 12). *The workplace-surveillance technology boom*. Slate Magazine. https://slate.com/technology/2020/05/workplace-surveillance-apps-coronavirus.html.

16. Cutter, S. L., Boruff, B. J., & Shirley, W. L. (2003). Social vulnerability to environmental hazards. *Social science quarterly, 84*(2), 242–261.

17. Dubois, E., Bright, D., & Laforce, S. (2021). Educating Minoritized Students in the United States During COVID-19: How Technology Can be Both the Problem and the Solution. *IT Professional, 23*(2), 12–18.

18. Duffy, R. (2021). 'Imagine how you would use teleportation': Drone delivery unicorn Zipline's CEO on plans for $250 million funding. Morning Brew. https://www.morningbrew.com/emerging-tech/stories/2021/06/30/imagine-use-teleportation-drone-delivery-unicorn-ziplines-ceo-plans-250-million-funding.

19. D'Urso, S. C. (2006). Who's watching us at work? Toward a structural–perceptual model of electronic monitoring and surveillance in organizations. Communication Theory (1050–3293), 16(3), 281–303. https://doi.org/10.1111/j.1468-2885.2006.00271.x.

20. eVisit. (2020). Telemedicine: ultimate guide - everything you need to know. https://evisit.com/resources/what-is-telemedicine/.

21. Eghtesadi, M. (2020). Breaking social isolation amidst COVID-19: a viewpoint on improving access to technology in long-term care facilities. *Journal of the American Geriatrics Society, 68*(5), 949.

22. Ezra, O., Cohen, A., Bronshtein, A., Gabbay, H., & Baruth, O. (2021). Equity factors during the COVID-19 pandemic: Difficulties in emergency remote teaching (ert) through online learning. *Education and Information Technologies*, 1–25.

23. Fox, S. (2000, August 20). *Trust and privacy online*. Pew Research Center: Internet, Science & Tech. https://www.pewresearch.org/internet/2000/08/20/trust-and-privacy-online/.

24. Frik, A., Nurgalieva, L., Bernd, J., Lee, J. S., Schaub, F., & Egelman, S. (2019). *Privacy and Security Threat Models and Mitigation Strategies of Older Adults*. 21.

25. Gevaert, C. M., Sliuzas, R., Persello, C., & Vosselman, G. (2018). Evaluating the societal impact of using drones to support urban upgrading projects. ISPRS international journal of geo-information, 7(3), 91.

26. Gilbert, D. (2020). China wants to use its coronavirus app to track how much its citizens sleep, drink, and smoke. https://www.vice.com/en_asia/article/v7gem3/china-wants-to-use-its-corona virus-app-to-track-how-much-its-citizens-sleep-drink-and-smoke.

27. GlobalDataHealthCare. (2020). Can wearable devices be used to predict Covid-19? Verdict Medical Devices. https://www.medicaldevice-network.com/comment/wearable-devices-covid-19/.

28. Grandia, L. (2014, 20 May). Healthcare Information Systems: Past, Present, Future. Health Catalyst. https://www.healthcatalyst.com/insights/healthcare-information-systems-past-present-fut ure/.

29. Greenaway, K. E., Chan, Y. E., & Crossler, R. E. (2015). Company information privacy orientation: A conceptual framework. Information Systems Journal, 25(6), 579–606. https://doi.org/10.1111/isj.12080.

30. Greenhalgh, T., Wherton, J., Shaw, S., & Morrison, C. (2020). Video consultations for covid-19: An opportunity in a crisis? BMJ 2020;368:m998 doi: https://doi.org/10.1136/bmj.m998 (Published 12 March 2020).

31. Greenwood, F. (2021, 30 July). Assessing the impact of drones in the global COVID response. Brookings. https://www.brookings.edu/techstream/assessing-the-impact-of-drones-in-the-glo bal-covid-response.

32. Health Catalyst. (2020). Health Catalyst Plans to Support Health System Clients' COVID-19 Response with Three Initial Solutions Focused on Patient Tracking, Public Health Surveillance and Staff Augmentation Support. https://www.healthcatalyst.com/news/health-catalyst-support-plan-to-support-health-system-clients-covid-19-response/.

33. Henderson, B. (2020). Drones to Deliver Protective Equipment in North Carolina. https://www.govtech.com/products/Drones-to-Deliver-Protective-Equipment-in-North-Carolina.html.

34. Holt, M., Lang, B., & Sutton, S. G. (2017). Potential Employees' Ethical Perceptions of Active Monitoring: The Dark Side of Data Analytics. Journal of Information Systems, 31(2), 107–124. https://doi.org/10.2308/isys-51580.

35. International Association of Privacy Professionals. (2020). *Privacy in the wake of COVID-19*. https://iapp.org/media/pdf/resource_center/iapp_ey_privacy_in_wake_of_covid_19_report.pdf.

36. Karmiyati, D., & Pradhan, D. (2021). Technology, Disasters, and Inclusivising Digital Access for Education in Society 5.0: Leading in the Borderless World. edited by Diah Karmiyati, Bildung Publishing.

37. Kaur, T., & Sharma, P. (2020). A study on working women and work from home amid coronavirus pandemic. *J Xi'an Univ Archit Technol*, 1400–1408.

38. Kehr, F., Kowatsch, T., Wentzel, D., & Fleisch, E. (2015). Blissfully ignorant: The effects of general privacy concerns, general institutional trust, and affect in the privacy calculus: Privacy calculus: dispositions and affect. Information Systems Journal, 25(6), 607–635. https://doi.org/10.1111/isj.12062.

39. Keith, T. (2020). Minn. Lawmakers consider limits on face-scanning technology for police. https://www.fox9.com/news/minn-lawmakers-consider-limits-on-face-scanning-technology-for-police.

40. Kimery, A. (2020, 15 May). London Metropolitan Police now considering pausing facial recognition expansion. https://www.biometricupdate.com/202005/london-metropolitan-police-now-considering-pausing-facial-recognition-expansion.

41. Landing AI. (2020). Landing A.I. Creates an AI Tool to Help Customers Monitor Social Distancing in the Workplace. https://landing.ai/landing-ai-creates-an-ai-tool-to-help-customers-monitor-social-distancing-in-the-workplace/.

42. Learned-Miller, E., Ordóñez, V., Morgenstern, J., & Buolamwini, J. (2020). Facial Recognition Technologies in the Wild.

43. Madianou, M. (2015). Digital inequality and second-order disasters: Social media in the Typhoon Haiyan recovery. *Social Media+ Society*, *1*(2), 2056305115603386.

44. Malhotra, N. K., Kim, S. S., & Agarwal, J. (2004). Internet users' information privacy concerns (IUIPC): The construct, the scale, and a causal model. *Information Systems Research*, *15*(4), 336–355. https://doi.org/10.1287/isre.1040.0032.

45. McFarland, K. (2020). How COVID-19 is driving a wave of innovation. https://washingtone chnology.com/articles/2020/04/16/insights-mcfarland-crisis-innovation.aspx.

46. Membrive, V., & Armie, M. (2020). Storytelling, Gamification, and Videogames: A Case Study to Teach English as a Second Language. Using Literature to Teach English as a Second Language (pp. 122–141). IGI Global.

47. Morris, A. (2020). Monitoring COVID-19 from hospital to home: First wearable device continuously tracks key symptoms. https://news.northwestern.edu/stories/2020/04/monitoring-covid-19-from-hospital-to-home-first-wearable-device-continuously-tracks-key-symptoms/.

48. Morrison, S. (2020, April 2). Just because you're working from home doesn't mean your boss isn't watching you. *Vox.* https://www.vox.com/recode/2020/4/2/21195584/coronavirus-remote-work-from-home-employee-monitoring.

49. Murphy, R., Manjunath, V. B., Gandudi, M., & Adams, J. (2020). *Robots are playing many roles in the coronavirus crisis—and offering lessons for future disasters.* https://www.govtech.com/products/Robots-Are-Playing-Many-Roles-in-the-Coronavirus-Crisis--and-Offering-Lessons-for-Future-Disasters.html.

50. Okazaki, S., Li, H., & Hirose, M. (2009). Consumer privacy concerns and preference for degree of regulatory control. Journal of Advertising, 38(4), 63–77. https://doi.org/10.2753/JOA0091-3367380405.

51. O'Neil, P. H., Ryan-Mosley, T., & Johnson, B. (2020). A flood of coronavirus apps are tracking us. Now it's time to keep track of them. MIT Technology Review. https://www.technologyre view.com/2020/05/07/1000961/launching-mittr-covid-tracing-tracker/.

52. Peek, S. T. M., Wouters, E. J. M., van Hoof, J., Luijkx, K. G., Boeije, H. R., & Vrijhoef, H. J. M. (2014). Factors influencing acceptance of technology for aging in place: A systematic review. *International Journal of Medical Informatics*, *83*(4), 235–248. https://doi.org/10.1016/j.ijmedinf.2014.01.004.

53. Pew Research Center. (2021). Online Privacy & Security—Research and data from the Pew Research Center. Pew Research Center. Retrieved September 16, 2021, from https://www.pew research.org/topic/internet-technology/technology-policy-issues/online-privacy-security/.

54. Pons, S. (2020). Morocco launches fleet of drones to tackle virus from the sky. https://finance. yahoo.com/news/morocco-launches-fleet-drones-tackle-virus-sky-014912077.html.

55. Pressgrove, J. (2020). Public Concern Grounds COVID-19 Drone Pilot in Connecticut. https:// www.govtech.com/products/Public-Concern-Grounds-COVID-19-Drone-Pilot-in-Connecticut. html.

56. Qureshi, M. (2021, April 8). *Facial recognition for COVID vaccination: What about data privacy?* The Quint. https://www.thequint.com/tech-and-auto/facial-authentication-for-covid-vac cination-is-this-a-viable-move.

57. Raji, I. D., Gebru, T., Mitchell, M., Buolamwini, J., Lee, J., & Denton, E. (2020, February). Saving face: Investigating the ethical concerns of facial recognition auditing. In Proceedings of the AAAI/ACM Conference on AI, Ethics, and Society (pp. 145–151).

58. Ratnam, G. (2020). Efforts to use ai in covid-19 research hit roadblocks. https://www.govtech.com/health/Efforts-to-Use-AI-in-COVID-19-Research-Hit-Roadblocks.html.
59. Rockwell, K., & Gilroy, A. (2020). Incorporating telemedicine as part of COVID-19 outbreak response systems. The American Journal of Managed Care, 26(4), 147–148. https://doi.org/10.37765/ajmc.2020.42784.
60. Saulnier, A., & Thompson, S. N. (2016). Police UAV use: Institutional realities and public perceptions. *Policing: An International Journal of Police Strategies & Management*.
61. Schwartz, S. (2020, August 20). As Teachers Livestream Classes, Privacy Issues Arise. *Education Week*. https://www.edweek.org/technology/as-teachers-livestream-classes-privacy-issues-arise/2020/08.
62. Schmidtlein, M. C., Deutsch, R. C., Piegorsch, W. W., & Cutter, S. L. (2008). A sensitivity analysis of the social vulnerability index. *Risk Analysis: An International Journal, 28*(4), 1099–1114.
63. Selinger, E., & Hartzog, W. (2020). The inconsentability of facial surveillance. Loy. L. Rev., 66, 33.
64. Shahroz, M., Ahmad, F., Younis, M. S., Ahmad, N., Kamel Boulos, M. N., Vinuesa, R., & Qadir, J. (2021). COVID-19 digital contact tracing applications and techniques: A review post initial deployments. *Transportation Engineering, 5*, 100072. https://doi.org/10.1016/j.treng.2021.100072.
65. Smith, H. J., & Hasnas, J. (1999). Ethics and information systems: the corporate domain. MIS Quarterly, 23(1), 109. https://doi.org/10.2307/249412.
66. Smith, H. J., Milberg, S. J., & Burke, S. J. (1996). Information Privacy: Measuring Individuals' Concerns About Organizational Practices. *MIS Quarterly, 20*(2), 167–196. https://doi.org/10.2307/249477.
67. Solon, O. (2018, February 1). Amazon patents wristband that tracks warehouse workers' movements. *The Guardian*. https://www.theguardian.com/technology/2018/jan/31/amazon-warehouse-wristband-tracking.
68. Soria, K. M., Chirikov, I., & Jones-White, D. (2020). The obstacles to remote learning for undergraduate, graduate, and professional students.
69. Stanton, J. M., & Stam, K. R. (2002). Information technology, privacy, and power within organizations: a view from boundary theory and social exchange perspectives. Surveillance & Society, 1(2), 152–190. https://doi.org/10.24908/ss.v1i2.3351.
70. Studentprivacymatters.org. (n.d.). Why students should be allowed to keep their cameras off during remote learning | Parent Coalition for Student *Privacy*. Retrieved September 16, 2021, from https://studentprivacymatters.org/how-students-should-be-protected-from-surveillance-during-remote-learning/.
71. Su, E. (2020, 8 May). Robot dogs are patrolling Singapore parks telling people to socially distance. Business Insider. https://www.businessinsider.com/roaming-robodog-politely-tells-singapore-park-goers-to-keep-apart-2020-5.
72. Surma, T., & Kirschner, P. A. (2020). Technology-enhanced distance learning should not forget how learning happens. *Computers in Human Behavior, 110*, 106390. https://doi.org/10.1016/j.chb.2020.106390.
73. Syakur, A. (2020). The effectiveness of English learning media through google classroom in higher education. *Britain International of Linguistics Arts and Education (BIoLAE) Journal, 2*(1), 475–483.
74. Tabora, V. (2020). *Applying Infrared Thermography For Coronavirus Screening*. https://medium.com/0xmachina/applying-infrared-thermography-for-coronavirus-screening-ee5b2dc8a6cb.

75. The Wire. (2021, April 14). *Digital rights bodies warn against use of facial recognition technology in vaccination drive*. The Wire. https://thewire.in/rights/covid-19-vaccination-facial-recognition-technology-aadhaar-vaccine.
76. Ting, D. S. W., Carin, L., Dzau, V., & Wong, T. Y. (2020). Digital technology and COVID-19. *Nature medicine, 26*(4), 459–461.
77. Tomlinson, S. B., Hendricks, B. K., & Cohen-Gadol, A. A. (2020). Editorial. Innovations in neurosurgical education during the COVID-19 pandemic: Is it time to reexamine our neurosurgical training models? *Journal of Neurosurgery*, 1–2. https://doi.org/10.3171/2020.4.JNS201012.
78. U.S. Government Accountability Office. (2021). *Facial Recognition Technology: Federal Law Enforcement Agencies Should Better Assess Privacy and Other Risks*. https://www.gao.gov/products/gao-21-518.
79. Verdict Medical Devices. (2020, 2 January). *What are the main types of robots used in healthcare?* https://www.medicaldevice-network.com/comment/what-are-the-main-types-of-robots-used-in-healthcare/.
80. Wagenseil, P. (2021, August 27). *Zoom security issues: Everything that's gone wrong (so far)*. Tom's Guide. https://www.tomsguide.com/news/zoom-security-privacy-woes.
81. Webster, P. (2020). Virtual health care in the era of COVID-19. *The Lancet, 395*(10231), 1180–1181. https://doi.org/10.1016/S0140-6736(20)30818-7.
82. Weiss, D. C. (2020). Another lawsuit is filed against Zoom over alleged privacy problems. *ABA Journal*. https://www.abajournal.com/news/article/another-lawsuit-is-filed-against-zoom-over-alleged-privacy-problems.
83. Weiss, D. C. (2021). Zoom agrees to $85M settlement in litigation over privacy and "Zoom-bombings." https://www.abajournal.com/news/article/zoom-agrees-to-85m-settlement-in-litigation-over-privacy-and-zoombombings.
84. Wells, J. (2020, May 19). Verint launches compliance recording for Microsoft Teams collaboration. https://www.kmworld.com/Articles/News/News/Verint-launches-compliance-recording-for-Microsoft-Teams-collaboration-140905.aspx.
85. World Health Organization. (2020). Ethical considerations to guide the use of digital proximity tracking technologies for COVID-19 contact tracing: interim guidance, 28 May 2020(No. WHO/2019-nCoV/Ethics_Contact_tracing_apps/2020.1). World Health Organization.
86. Xie, B., He, D., Mercer, T., Wang, Y., Wu, D., Fleischmann, K. R., Zhang, Y., Yoder, L. H., Stephens, K. K., Mackert, M., & Lee, M. K. (2020). Global health crises are also information crises: A call to action. Journal of the Association for Information Science and Technology (JASIST) https://doi.org/10.1002/asi.24357.
87. Xie, B., Charness, N., Fingerman, K., Kaye, J., Kim, M. T., & Khurshid, A. (2020). When going digital becomes a necessity: Ensuring older adults' needs for information, services, and social inclusion during COVID-19. *Journal of Aging & Social Policy, 32*(4–5), 460–470.
88. Yu, E. (2020). Singapore looking at wearable devices to support COVID-19 contact tracing. ZDNet. https://www.zdnet.com/article/singapore-looking-at-wearable-devices-to-support-covid-19-contact-tracing/.
89. Yun Technology. (2020). *QuickDoctorOnline*. https://www.120ask.com/.

Thora Knight Thora Knight is an attorney in the privacy and cybersecurity group at an Am Law 200 firm. She has strong practical and academic experience in information security, privacy, technology, and intellectual property. Thora assists clients in implementing strategies to assess and mitigate cybersecurity and privacy risks, respond to cyber-attacks and data breaches, and maintain compliance with federal, state, and foreign privacy and data protection laws.

In the academic setting, Thora's broad research interests lie at the intersection of privacy, security and emerging technologies. Thora has analyzed and applied privacy and cybersecurity standards including the National Institute of Standards and Technology Cybersecurity Frameworks and Federal Trade Commission and Department of Homeland Security best practices for securing connected networks to educational training. Thora is pursuing a Ph.D. in Information Science at the University at Albany, focusing on the regulatory effects of privacy legislation on behavioral advertising. She earned her JD and MBA from the State University of New York at Buffalo and holds a BS in business with an emphasis on information systems from the University of Phoenix.

Xiaojun Yuan, Ph.D., is an Associate Professor in the College of Emergency Preparedness, Homeland Security, and Cybersecurity at the University at Albany, State University of New York. Her research interests include both Human Computer Interaction and Information Retrieval, with the focus on user interface design and evaluation and human information behavior.

She has received various grants and contracts, including from the Institute of Museum and Library Servcices, SUNY seed grant, Initiatives For Women Program at University at Albany, and New York State Education Department.

She published extensively in journals in information retrieval and human computer interaction (JASIS&T, IP&M, Journal of Documentation, etc.), and conferences in computer science and information science (ACM SIGIR, ACM SIGCHI, ACM CHIIR, ASIS&T, etc.).

Dr. Yuan received her Ph.D. from Rutgers University at the School of Communication and Information, and Ph.D. from Chinese Academy of Sciences in the Institute of Computing Technology. She received her M.S. in Statistics from Rutgers University and M.E. and B.E. in Computer Application from Xi'an University of Science & Technology in China. She serves as an Editorial Board Member of Aslib Journal of Information Management (AJIM), and a Board Member of the International Chinese Association of Human Computer Interaction. She is a member of Association for Information Science and Technology (ASIS&T), the Association for Computing Machinery (ACM) and the Institute of Electrical and Electronics Engineers (IEEE).

DeeDee Bennett Gayle, Ph.D. is an Associate Professor in the College of Emergency Preparedness, Homeland Security, and Cybersecurity at the University at Albany, State University of New York. She broadly examines the influence and integration of advanced technologies on the practice of emergency management, and for use by vulnerable populations.

She has secured research grants and contracts, including from the National Science Foundation, Federal Emergency Management Agency, and the Department of Homeland Security. Her work is published in various journals, and she has presented at several conferences related to emergency management, disability, wireless technology, and future studies.

Dr. Bennett Gayle received her Ph.D. from Oklahoma State University in Fire and Emergency Management. She has a unique academic background having received both her M.S. in Public Policy and B.S. in Electrical Engineering from the Georgia Institute of Technology. She is an Advisory Board Member for the Institute for Diversity and Inclusion in Emergency Management (I-DIEM), a member of the Social Science Extreme Events Reconnaissance (SSEER) and Interdisciplinary Science Extreme Events Reconnaissance (ISEER), within the NSF-FUNDED CONVERGE initiative.

Salimah LaForce, M.S. is a research scientist II at the Georgia Institute of Technology and senior policy analyst at Georgia Tech's Center for Advanced Communications Policy. She specializes in policy research, identifying and describing intended and unanticipated implementation outcomes.

Her work spans a variety of topic areas, including increasing accessibility and usability of information communications technologies, improving employment outcomes for individuals with disabilities, building capacity for inclusive emergency response efforts, and cultural competency in the delivery of general and mental healthcare services. She has 15 years of experience conducting user needs and experiences research and utilizing study results to inform policy and practice recommendations.

Salimah earned her B.A. in English literature from Agnes Scott College and her M.S. in Clinical Psychology, applied research specialization, from the Harold Abel School of Social and Behavioral Sciences, Capella University. Her graduate studies focused on culturally competent delivery of mental health services and the inclusion of people with disabilities, ethnic minorities, **and women** in the workplace with an emphasis on the function of perceptions, bias, and social attitudes.

Conclusion

8

Dan Wu and Le Ma

Abstract

The health risks of socially vulnerable groups, such as the elderly, the sick, and the disabled, are significantly elevated under the COVID-19 epidemic. Therefore, the different factors affecting the use of information technology by socially vulnerable groups under COVID-19 are explored at the level of the use of emerging information technology. The impact on the information behavior of socially vulnerable groups under COVID-19 is also explored at the level of information behavior, including health information needs, the digital divide phenomenon, and the utilization of public information services. Based on the above findings, the current status of information behavior research for socially vulnerable groups is combined. Future research directions of information technology and information behavior for socially vulnerable groups are proposed. First, to improve the research theory of information behavior of socially vulnerable groups regarding information technology. Second, to apply big data technology and data analysis technology to explore the information technology adoption behavior of socially vulnerable groups in-depth. Third, to construct the information behavior model of socially vulnerable groups based on empirical research cases. Fourth, to use information technology for socially vulnerable groups according to information technology and the barriers faced by socially vulnerable groups in using information technology, and to provide strategies for using information technology that meet the needs of socially vulnerable groups.

D. Wu (✉) · L. Ma
School of Information Management, Wuhan University, Wuhan, China
e-mail: woodan@whu.edu.cn

Keywords

COVID-19 • Socially vulnerable group • Information technology • Information behavior

During the COVID-19 pandemic, due to the COVID-19 epidemic prevention and control policies of various countries around the world, such as social isolation, school shutdown, and border closures, people who were not previously vulnerable may become vulnerable when they suddenly lose their economic income or social support, and it is difficult to determine which groups will become vulnerable. In the context of the COVID-19 pandemic, vulnerable groups in society include not only the elderly with ill health and complications of the disease and those who are homeless or have no fixed home, but also people from all socioeconomic groups who are struggling to cope with economic, mental or physical crises [3]. The current phenomenon of vulnerable groups and health inequalities in society is becoming increasingly evident as COVID-19 spreads across the globe.

In response to the COVID-19 pandemic, the risk of the virus spreading to vulnerable groups is increasing. The elderly, the sick, people in long-term care, people living in densely populated urban areas, and people living in rural areas with limited access to health care are all threatened by the COVID-19 outbreak. In recent years, the rapid development of information technology provides technical support for socially vulnerable groups to realize the relative fairness of information. Through the combination of information technology and information behavior of socially vulnerable groups, on the one hand, helps to build an information service system suitable for socially vulnerable groups under public health emergencies. On the other hand, it is beneficial to improve the information asymmetry of vulnerable groups in public health emergencies. Book by scientific and objective research methods to explore COVID-19 outbreak, technology adoption to the effects of the information behavior of the socially vulnerable groups, and from the reality of socially vulnerable groups, to explore COVID-19 outbreak of socially vulnerable groups under information behavior characteristics, information awareness, information elements such as dealing with behavior, found and summing rules, requirements and deficiencies, In order to provide theoretical basis and decision-making support for the prevention and response of public health emergencies in the future.

8.1 Research Summary

8.1.1 The Impact of the Adoption of Emerging Information Technologies on Vulnerable Groups in Society During the Covid-19 Pandemic

(1) Research on the use of information technology by vulnerable groups in the context of COVID-19 pandemic.

The wide application of information technology and information system has become one of the characteristics of the information age, and the adoption and utilization of information technology are one of the research hotspots in the field of information systems. Explore the need for focusing on technology adoption during the covid-19 pandemic, especially in underserved communities. Based on the Bronfenbrenner framework, this paper explores the impact of the use of information technology at the levels of individuals, communities, organizations, and societies. Based on the Bronfenbrenner framework, a series of problems arising from the use of information technology by the socially marginalized population are found. On the one hand, the research results provide a theoretical basis for further research on the impact of information technology on socially vulnerable groups. On the other hand, it provides a research direction for other scholars and practitioners to understand the impact of information technology use on expanding or reducing the vulnerability of the marginalized population.

(2) Research on the factors influencing the use of technology by socially vulnerable groups during COVID-19 pandemic.

During the COVID-19 global pandemic, there was significant inequality in the use of information technology by socially vulnerable groups. Too much information (including correct, incorrect, or false) makes it difficult for people to find trustworthy sources of information and do not know how to deal with such information independently. Therefore, by systematically summarizing the factors in the process of using artificial intelligence technology through the method of literature review, which factors will affect the use of artificial intelligence technology by socially vulnerable groups, the motivation of socially vulnerable groups to adopt artificial intelligence technology, and the specific application of artificial intelligence technology. This research helps to improve the design of artificial intelligence systems for socially vulnerable groups and helps solve the ethical problems in using artificial intelligence systems for socially vulnerable groups. At the same time, when discussing the implications of rapid technology adoption, privacy issues are also considered.

8.1.2   Research on the Information Behavior of Socially Vulnerable Groups During the Covid-19 Pandemic

(1) Research on the information behavior of different socially vulnerable groups in the context of COVID-19.

To explore the characteristics and evolution of health information needs of COVID-19 patients. Through the collection and analysis of question and answer (Q&A) data from social Q&A communities, it is found that the health information needs of COVID-19 patients are diverse, which can be divided into five categories: C1 etiology, symptoms and manifestations, C2 examination, and diagnosis, C3 prevention, C4 treatment, and C5 infectivity. The study also found different types of health information needs of COVID-19 patients in different periods of the COVID-19 outbreak, the unique health information needs of COVID-19 patients are infectious, and the health information needs COVID-19 patients with different social attributes are different. This study clarified the health information content required by COVID-19 patients in different periods of COVID-19 outbreak and provided a research basis for further optimizing the information display mechanism and information resource organization of socially vulnerable groups in the social Q&A community. In the study of factors affecting elderly access to information during the COVID-19 epidemic, a semi-structured interview method combined with a content analysis method was adopted to identify the digital divide that older people will face during the COVID-19 pandemic. The study identifies the digital divide the elderly face during the COVID-19 pandemic. It is found that the digital divide dilemma faced by the elderly in the COVID-19 era mainly includes two aspects: assessing the authenticity of information and accessing digital services. The reasons for the digital divide include personal factors and external factors. Besides, suggestions are put forward in response to the digital divide dilemma faced by the elderly during the COVID-19 pandemic. For information technology(IT)enterprises, appropriate interfaces and interactive content should be designed according to the usage habits and characteristics of the elderly. For public service organizations, technical solutions for the elderly are provided. The research aims to address better the digital divide older adults face during the COVID-19 pandemic so that they have faster access to accurate information to protect their health better.

(2) Factors contributing to the widening of the digital divide among vulnerable groups in the COVID-19 pandemic.

During the COVID-19 pandemic, countries have issued prevention and control policies to avoid social contact. As a result of the epidemic prevention and control policies, socially vulnerable groups with limited literacy skills, lack of computer knowledge, limited wi-fi access are increasingly experiencing inequality in access to the internet, navigation of websites, and online registrations to gain access to benefits, the court system, and medical care. Furthermore, individuals who were already within the access to justice gap became even more marginalized when legal systems and public benefits went virtual. Therefore, by studying the reasons for the expansion of the digital divide faced by the vulnerable groups in society, we can prevent the vulnerable groups in society from becoming more vulnerable in the legal system and public welfare. This research will help clarify the specific content of technological barriers faced by vulnerable groups in the COVID-19 pandemic and provide solutions to effectively avoid further widening the digital divide among vulnerable groups in society in the future.

(3) Research on barriers of socially vulnerable groups using public information services during COVID-19.

By dividing the socially disadvantaged into two broad categories, the first kind of vulnerable groups are those who lack the corresponding ability and are threatened by epidemic because of their physical conditions or epidemic prevention measures, including the disabled, the visually impaired, the hard of hearing. The second kind of vulnerable group is the group in which individuals are in a nasty pasty environment, which makes them vulnerable to the threat of virus, or lose their financial resources because of prevention and control measures, including the Frontline medical workers, couriers, and the unemployed. Found that different socially vulnerable groups in the use of public information services face the difficulty of the specific contents and differences. Furthermore, according to the different socially vulnerable groups to use the plight of public information service, from the policies and regulations, information technology and information system, information content, the management mechanism, operating mechanism, put forward the corresponding improvement Suggestions. This research helps other scholars understand the difficulties socially disadvantaged groups face in accessing public information services during the COVID-19 pandemic and provides targeted solutions for different socially disadvantaged groups.

8.2 Research Prospect

As the COVID-19 pandemic continues to roll back and forth, the number of marginalized socially vulnerable groups in society increases. Some researchers have made some progress in the research field of user information behavior based on information technology, but the research of information behavior for socially vulnerable groups is still in the exploratory stage. Because the research on the information behavior of socially vulnerable groups is insufficient, it is necessary to expand further the research on the information behavior of socially vulnerable groups based on information technology. The problems existing in the current research cannot be ignored. Firstly, the lack of new research topics and directions leads to the current research content is not rich enough. Secondly, there is a lack of systematic information behavior theoretical research framework, and the theoretical content is not paid enough attention. The analysis method is not standardized enough. There are practical problems such as difficulty collecting data samples, the threshold of promotion, and application. Third, the research perspective of the information behavior of socially vulnerable groups based on information technology is limited, and the intervention of new information technology is lacking, which requires the support of new information technology. Combined with previous research prospects and the evolution and development of information technology, this book proposes that future research in this field should focus on the following aspects.

8.2.1 Improve Basic Theoretical Research

A comprehensive analysis of current research shows that the theoretical framework of information behavior research of socially vulnerable groups based on information technology is not perfect, and the research topic in this field is too focused, which mainly focuses on a specific information behavior analysis, without considering that the whole information behavior is a continuous process. Furthermore, each part is an interactive, overlapping influence. The existing research lacks comprehensive thinking and does not conduct a holistic study on information behavior. In addition, information technology, as a technology to study how people obtain, process, transmit and use information, can affect the behavior of socially vulnerable groups to obtain, disseminate and use information, and thus help realize the relative equality of information utilization. Information technology provides new ideas and perspectives for exploring the information behavior of vulnerable groups. In the studies on groups, previous studies have carried out internal analysis of information behavior for one or two different characteristics of groups, but lack of multi-dimensional and multi-level consideration of the characteristics and influencing factors of information behavior from the aspects of the region, occupation, economic status, and living environment. Therefore the future research can revolve around the information behavior of socially vulnerable groups, from the perspective of the user experience of information technology research and analysis information behavior characteristics of socially vulnerable groups, analyzing an information technology or several functions, interface, interaction between information technology, navigation, and the difference of content attribute to the influence of the socially vulnerable groups information behavior. At present, the research content of information behavior of socially vulnerable groups based on information technology mainly comes from research on technology adoption behavior, information behavior characteristics, and the digital divide. Since the COVID-19 epidemic is sudden, socially vulnerable group's information technology behavior change is not entirely of cognition, and information fair judgment standard is relative, dynamic. In addition, information fairness is not to deny the objective existence of information differences but to make these differences reach an effective balance to form a harmonious and orderly system structure. It can be seen that the construction of basic theories related to information technology behavior research is insufficient as a whole, and only a few exploratory theoretical studies have been conducted. Bhattacherjee [1], for example, proposed that expectation confirmation theory (ECT) is one of the most widely used models in the study of continuous use of information technology, which reveals how information technology affects users' intention to continue using by describing cognitive psychology and cognitive behavior. Venkatesh and Bala [4] proposed TAM3 model began to focus on the question of the long-term use of information technology. This model thinks in the adoption of information technology environment, experience is one of the important regulating variable information technology adoptions, the user's response to information technology will change with time, and constantly changing perception in determining the user continues to play an important role in the process of using information technology. Chang [2] put forward the theory related to vulnerable groups. He defined vulnerable groups under the outbreak from two dimensions and proposed that different kinds of vulnerable need special protection under the epidemic prevention and control.

8.2.2 Application of Big Data Technology and Data Analysis Technology

At present, with the wide application of the new generation of information technology with big data, cloud computing, Internet of Things, virtual reality, and artificial intelligence as the core. These emerging information technologies constantly promote the development and evolution of user information behavior, and users' intentions and related cognition in the process of information technology adoption is not invariable, which will change the user's behavior of using technology. In studying user's information behavior, new information technology can help users obtain and use information, but also can greatly improve the way users obtain and use information, which can become one of the effective ways to improve user's information behavior. It can be seen that emerging information technology can provide important technical support for studying the information behavior of socially vulnerable groups under the COVID-19 pandemic. A lot of basic data can be collected through these information technologies. These data become the research samples of information technology behavior of socially vulnerable groups. Improving the efficiency of information technology behavior research of socially vulnerable groups through data analysis has become one of the important research directions in this field in the future. At the same time, research on the entire process of information technology adoption by socially disadvantaged groups will also become one of the research hotspots in information technology research. In the big data environment, different basic data forms can be used to show the behavior of socially disadvantaged groups in specific situations. The use of data mining and data visualization technologies will help to explore the information contained in the data at a deeper level. Therefore, it will provide important technical support for the future study of information technology behavior of socially vulnerable groups.

8.2.3 Construction of Behavior Model Based on Empirical Case Study

During the COVID-19 epidemic, the research on the information behavior of socially vulnerable groups based on information technology aims to solve the problems socially vulnerable groups face, such as technology adoption, digital divide, and identification of false information. In addition, by analyzing the information behavior rules of socially vulnerable groups under the information technology environment, to address the information inequality faced by vulnerable groups in society during the COVID-19 pandemic. However, the research on the information behavior of socially vulnerable groups based on information technology involves a series of systematic and complex problems. From the perspective of socially vulnerable groups, the information behavior research of socially vulnerable groups based on information technology needs to face complex groups involving different ages, economic levels, and situations of the population. The research of information behavior mainly includes three levels. The first level is the user's level, which needs to borrow the viewpoints and methods of psychology and cognitive science to study each user's response to different information from the cognitive perspective. The second level is the group level: it usually borrows some theories and methods in the field of sociology, such as the daily life information behavior model, to study user behavior from the group perspective, that is, how users interact with other users in the group and solve information problems in daily life. The diversity of research subjects and the complexity of individual behavior characteristics of socially vulnerable groups determine that the realization of information equity is a complex research process. The third level is cultural aspects: theories and methods, such as community informatics, are usually borrowed from anthropology to study the cultural influence of user information behavior. Based on this, empirical methods should be widely used in the research process of information technology behavior of socially vulnerable groups. It is an important development direction of information technology behavior research on socially vulnerable groups in the future to construct the corresponding information behavior model through empirical research methods of data collection and analysis of specific cases of socially vulnerable groups to provide support for future decision making and improve the pertinence and substantiality of research.

8.2.4 Strategy Research of Information Technology Behavior Rule Based on Socially Vulnerable Groups

The information overload during the COVID-19 pandemic is beyond people's ability to process and understand information, and there is a large amount of false information and unverified misinformation in the vast amount of information on COVID-19. Such information can mislead people into making wrong decisions and actions and cause unnecessary physical and psychological harm. People's information decision guides their information behavior as the bottom basis of decision making. However, socially vulnerable groups have obvious disadvantages in information acquisition, screening, analysis, understanding, and application due to their limited ability to use information technology. Therefore, the purpose of exploring the information technology behavior of socially vulnerable groups during the COVID-19 epidemic is to provide decision-making support for addressing the practical problems of inadequate technology use capacity, digital divide, and information inequality of socially vulnerable groups in the context of major health emergencies. By summarizing the information technology behavior rules of socially vulnerable groups under major health emergencies, it can provide better guidance for socially vulnerable groups in terms of information acquisition, information sharing, and information utilization. Therefore, this research issue is also one of the important contents of future research.

References

1. Bhattacherjee, A. (2001). Understanding information systems continuance: An expectation-confirmation model. *MIS quarterly*, 351–370. DOI: 10.2307/3250921
2. Chang, J. (2020). Special Protection of the Human Rights of the Four Vulnerable Groups Under the Sudden Major Epidemic Situation. *Journal of Human Rights*, 2020(01), 5–12. DOI: CNKI: SUN: RQYJ.0.2020-01-003
3. Lancet, T. (2020). Redefining vulnerability in the era of COVID-19. *Lancet (London, England)*, 395(10230), 1089. DOI: https://doi.org/10.1016/S0140-6736(20)30757-1
4. Venkatesh, V., & Bala, H. (2008). Technology acceptance model 3 and a research agenda on interventions. *Decision sciences*, 39(2), 273–315. DOI: https://doi.org/10.1111/j.1540-5915.2008.001 92.x

Dan Wu, Ph.D., is a Professor in the School of Information Management at Wuhan University, a member of the Academic Committee of Wuhan University, and the director of the Human-Computer Interaction and User Behavior Research Center. Her research areas include information organization and retrieval, user information behavior, human-computer interaction, and digital libraries.

She has secured research grants and contracts, the National Social Science Foundation of China (Major Program), the National Natural Science Foundation of China (NSFC), and the Humanities and Social Sciences Foundation of the Ministry of Education.

Her work is published in various journals (including the IP&M, JASIST, etc.). She has presented at several conferences related to information behavior, information retrieval, and digital library (SIGIR, CIKM, CSCW, CHIIR, etc.).

Dr. Wu received her Ph.D. from the Peking University in Management. She serves as the Editor-in-Chief of Aslib Journal of Information Management and the Executive Editor of Data and Information Management. She also serves as a director at large of the ASIS&T. She is a member of the ACM Digital Library Committee, a member of the iSchool Data Science Curriculum Committee, and a member of the JCDL Steering Committee.

Le Ma is a Ph.D. candidate at the School of Information Management, Wuhan University. Her research interests include information retrieval, user information behavior, etc. She is a member of the Human-Computer Interaction and User Behavior Research Center of Wuhan University. She participated in several Chinese national and provincial research projects and published several academic papers during her doctoral study.